THE POLITICS OF ENERGY RESEARCH AND DEVELOPMENT

Energy Policy Studies
Volume 3

THE POLITICS OF ENERGY RESEARCH AND DEVELOPMENT

Energy Policy Studies
Volume 3

Edited by

John Byrne and Daniel Rich

Routledge
Taylor & Francis Group
LONDON AND NEW YORK

First published 1986 by Transaction Publishers

Published 2017 by Routledge
2 Park Square, Milton Park, Abingdon, Oxon OX14 4RN
711 Third Avenue, New York, NY 10017, USA

First issued in hardback 2017

Routledge is an imprint of the Taylor & Francis Group, an informa business

Library of Congress Catalog Number: 86-1307

Library of Congress Cataloging in Publication Data

Main entry under title:

The Politics of energy R&D.

(Energy policy studies ; v. 3)
Bibliography: p.
1. Energy policy—United States—Addresses, essays, lectures. 2. Power resources—Research—United States—Addresses, essays, lectures. 3. Energy development—United States—Addresses, essays, lectures. I. Byrne, John, 1949- . II. Rich, Daniel. III. Title: Politics of energy R&D. IV. Series.
HD9502.U52P63 1986 333.79'0973 86-1307
ISBN 0-88738-653-9

ISSN: 0882-3537
ISBN 13: 978-1-138-53753-8 (hbk)
ISBN 13: 978-0-88738-653-4 (pbk)

For Lillian

and in Memory of Max

ACKNOWLEDGEMENT

The editors wish to thank W. Michael McCabe, former staff director for the Energy Conservation and Power Subcommittee of the Energy and Commerce Committee, U.S. House of Representatives, for his considerable assistance in the preparation of this book. His comments, criticisms and suggestions during the project were extremely valuable.

We also wish to express our gratitude to Patricia Grimes for preparing the manuscript with extraordinary skill, good humor and patience.

CONTENTS

Introduction: The Tragedy of U.S. Energy R&D Policy

Richard L. Ottinger

The tragedy of federal energy research and development policy over the past two decades is that it has been formulated totally without regard to our national energy needs.

Ideology, technological availability and industrial bias have determined energy R&D policy, virtually without thought of whether the policy satisfies our national social and economic needs. Under the Reagan administration the biases are so strong that the signals of the marketplace, ostensibly the Reagan rallying cry, are blatantly ignored in favor of subsidization of nuclear power technology which the marketplace has resoundingly rejected.

While under the Carter administration much greater balance was achieved - conservation and renewable energy resources, the most economical and urgently needed options, were promoted - R&D was heavily tilted to central station electric generation in the face of national priorities calling for technological advances in fields that could relieve our dependence on foreign oil.

The startling fact is that in the mid-1970s only 10 percent of our oil was used to generate electricity, while more than 52 percent was used for transportation and 26 percent for industrial use. Today less than 5 percent of our oil is used for electricity, yet two-thirds of the President's fiscal year 1986 energy technologies R&D budget request is devoted to nuclear fission and fusion programs which would be used for central station electric generation.

This anomaly is accounted for in part by a fascination in the scientific community with high technology options on the threshold of modern science. Solving the mysteries of the atom is sexy and academically rewarding, while the more mundane conservation and renewable technologies are accorded much less prominence in scientific academia.

Then, too, there is a strong political penchant for the quick fix. It is much easier for politicians and industry to visualize building gargantuan machines to provide new energy supplies than it is to visualize how to influence millions of individuals and thousands of businesses to adopt conservation practices like purchasing more efficient light bulbs, refrigerators and air conditioners or insulating roofs and walls. It is also politically more rewarding to finance big production units with their influx of megabucks and jobs than it is to achieve conservation.

Lastly, our federal energy R&D imbalance is accounted for by the greed of big industry and big labor whose eyes light up at billion dollar construction contracts - and by the simply incredible stupidity of the utility industry in pushing plants requiring huge capital investment and long lead times, plants that are uneconomical, even to the extent of bankrupting them in the process - while largely ignoring low cost, high cash-flow investments that are so economically attractive that independent firms are proliferating to take advantage of the opportunities the utilities eschew.

More amazing yet, the handwriting was clearly on the wall. Study after study, many sponsored by the government agencies which ignored them, clearly pointed to giving priority to conservation and renewables as the cheapest, cleanest, quickest means to achieve our real national objectives of economic energy independence. Starting with the Ford Foundation study, "A Time to Choose," in 1974, and confirmed by such establishment studies done by Roger Sant at Carnegie Mellon in 1979 and Robert Stobaugh and Daniel Yergin at the Harvard Business School in 1979, and reinforced by numerous government-sponsored General Accounting Office and National Academy of Science studies, the economic advantages of a conservation and renewables strategy and its superiority in achieving national energy independence was amplified time and time again.

Most fascinating of all, the Department of Energy's (DOE) own Solar Energy Research Institute study, "A New Prosperity: Building a Sustainable Energy Future," developed in 1980 and buried by the Reagan administration in 1981, detailed the available technologies, savings achievable and cost economies in each sector of energy use, while the department itself continued to push precisely the opposite priorities with hardly a mention of its own or the other studies.

In 1983, three years after the administration took office, a new study was released by DOE which tried to reverse economic trends and set the stage for a new central plant construction campaign. The study, "The Future of Electric Power in America," hailed by the Edison Electric Institute as a long overdue energy policy reorientation, totally ignored conservation because DOE said its contribution was too difficult to compute. Yet the study claimed flabbergasting new electric energy construction requirements through the year 2000 of 438 gigawatts at a cost of $1.8 trillion (over $1 trillion in constant 1982 dollars) - the equivalent of two-thirds of all U.S. savings, three times the existing net asset value of all our utilities, $14,000 per every U.S. household!

While this study has been largely discredited, DOE is still trying to fulfill these outlandish predictions with every imaginable subsidy for construction of new central station power plants - tax subsidies, new R&D commitments, automatic Construction-Work-In-Progress payments, continued Price-Anderson insurance subsidies - that in total account for a large portion of the initial cost of a new reactor, and they are still uneconomic.

The Reagan administration is also pushing hard to promote use of the full nuclear fuel cycle, with breeder reactors and reprocessing that are orders of magnitude uneconomic, while ignoring R&D on more efficient reactors and fuel burn-up technologies which together could extend fuel supplies indefinitely and vastly improve nuclear economics - to say nothing of R&D on much more economic conservation and renewables.

The Department of Energy and nuclear industry seek to mislead the Congress and public by claiming that electric generation is directly related to energy independence, as if one could economically use electricity to power our transportation fleet, the principle consumer of our oil supplies.

Amory and Hunter Lovins have done the pioneering studies to show the advantages of the "soft path" conservation, cogeneration and renewables options and the follies of pursuing central station electricity, particularly the most vulnerable, cumbersome and uneconomical nuclear options. In his chapter for *The Politics of Energy R&D,* Amory Lovins underscores the startling facts that since 1979 the U.S. has obtained more than 100 times more new energy from conservation than from all other supply options, more renewable energy than nonrenewable, indeed, more small hydro and wind power than coal and nuclear energy combined. He also indicates the enormous potential of new conservation and renewables R&D in such areas as more efficient appliances, light bulbs, photocopiers, computers, smelters and electric motors. Lovins points out that the recognition of these opportunities has been and still is obstructed by a prevailing high-tech, hard-tech mindset; a mindset that has placed its hopes in nuclear power and, for reasons he examines in detail, has delivered instead a nuclear fiasco.

Richard T. Sylves' chapter on the rise and fall of Project Plowshare provides a vivid description of both the public and scientific optimism about nuclear energy that typified the post World War II era and the subsequent erosion of optimism as the technological, economic and political liabilities of nuclear power became apparent. During the 1950s, pronouncements concerning the peaceful uses of nuclear power promised nothing less than an energy utopia. Even so, as Sylves points out, Project Plowshare was surely a case of nuclear exotics; under the tutelage of the AEC, the U.S. government set out on a program to use nuclear bombs to dig canals and harbors, open up new mining deposits, and level the landscape as needed for transportation and commerce. Sylves catalogs the projects contemplated, planned, and sometimes carried out under the banner of Plowshare and describes the conditions that led to Plowshare's decline and finally its termination.

Eugene Frankel, the superb conservation and renewables expert formerly with the U.S. House Science and Technology Committee (on which I served), describes the competing policy directions that guided the expenditure of nearly $3 billion for solar R&D during the decade from 1974-83. In his analysis, Frankel points to particularly promising areas of R&D, such as wind machines, photovoltaics and eventually separation of hydrogen from water using solar technologies. He demonstrates, however, what I can testify to from experience: approaches to solar R&D in successive periods were shaped as much by politics and ideology as by technological opportunities. Indeed, he argues that it is only through examination of the interaction of technology, politics and ideology that we can come to a viable understanding of the government R&D process and of our energy R&D options.

In her chapter, Maxine Savitz, who supervised conservation and renewable programs within the Department of Energy, demonstrates the great potential of R&D in building construction, industrial processes, and cogeneration - and the impressive strides that were made in these areas under her tutelage. She argues that an active federal role in conservation R&D is appropriate and demonstrably cost-effective and successful. Moreover, the possibilities for even greater contributions to energy efficiency in the future are substantial and documentable. For her good work, Savitz was summarily (and I contend illegally) dismissed by the Reagan administration - a tragic loss to government of one of its most dedicated and talented civil servants (the circumstances of her dismissal are reviewed in a case study prepared at the Kennedy School of Government, Harvard University, entitled "Dr. Savitz and the Department of Energy").

Federal government programs to accelerate the introduction of new energy technologies into the marketplace have come under considerable attack from the Reagan administration which has sought to reduce the government's role in the commercialization process. J. David Roessner points out that despite such objections, the confluence of domestic and international forces that shape energy options will continue to call for solutions that involve government efforts to facilitate energy technology development. Roessner reviews the commercialization issues that need to be addressed in energy technology policy. Drawing upon case studies of solar energy commercialization strategies, he points to a variety of mismatches between the energy technologies and the commercialization strategies adopted in U.S. R&D policy. As a result, commercialization programs have sometimes been less successful than they might have been and than they may be in the future if we can learn from experience.

Energy R&D policies have been shaped by values and expectations that extend beyond immediate policy considerations. Identification of three of these influences is the subject of the chapter by John Byrne and Daniel Rich. First, they describe the evolution of a general R&D policy orientation in the United States that drew its inspiration and a good deal of its institutional apparatus from military R&D programs in World War II, especially the Manhattan Project. Second, they point to a prevailing understanding of energy change as based upon a number of energy myths and an underlying expectation of energy abundance brought about by one or another technological fix. Third, they consider the influence of technological values on energy choice and argue that it has sustained a consistent preference for large-scale, centralized technologies and an undervaluing of the social and political dimensions of energy choice.

Finally, Grant Thompson places the discussion of energy R&D into the context of national energy policy and reminds us that a number of key issues placed on the policy agenda in the 1970s have yet to be resolved. In Thompson's view, we have made some real progress in coming to grips with some of the challenges posed by the energy crisis but we face the danger of becoming quiescent when there is still a great deal that must be done. Among the items high on his agenda for action are the need to redress the burden on the poor created by higher energy prices, the need to pursue a more aggressive program of energy conservation, and the need to clarify public and private responsibilities in research and development and then strengthen the R&D process. Whether or not one agrees with the specifics of Thompson's agenda, it is difficult to quarrel with his analysis that in recent years we have walked away from any coherent energy policy while "muttering the mantra of free market," or with his warning that we can't afford to wait for the next shoe to drop.

The Politics of Energy R&D is essential reading for anyone who wants to understand the incredible tragedy of failed opportunities to permanently relieve our dependence on imported oil and to understand the waste of billions of taxpayers' dollars on uneconomic central station electric power options. The book also makes an important contribution in explaining the huge potential of more sound energy R&D alternatives - and the total failure of the Reagan administration to learn anything from the mistakes of the past or to develop the great potential of conservation and renewables for the future.

The Origins of the Nuclear Power Fiasco

Amory B. Lovins

The federal commitment to civilian nuclear power has achieved six things which are individually noteworthy and collectively extraordinary.

First, it has cost the taxpayer upwards of $70 billion (1983 dollars) in direct federal support and subsidy to bring the technology to a state where it can be commercially deployed and to bring that deployment to its present state of disarray (updated from Battelle Pacific Northwest Laboratories, 1978).

Second, the federal commitment has led the electric utility industry to allocate great sums of capital to reactor construction - over $125 billion so far, which is more than the entire cost of the space program or the war in Vietnam (Cook, 1985). In 1982 alone, utilities spent twice as much money building reactors as the nation invested in the car, truck, iron and steel industries put together.

Third, the nuclear commitment has slowed the resolution of the energy problem more than has any other federal action, with the possible exception of price regulation on natural gas. This is because money spent on slow and costly ways of replacing oil, like building power plants, cannot be spent on rapid and relatively cheap ways of saving oil, like weatherizing buildings or making vehicles more efficient.

Fourth, nuclear power has gravely undermined national security (on which another branch of our government is spending about $10,000 a second), both by spreading bombs and innocent disguises for bomb programs around the world (Lovins and Lovins, 1981), and by putting very large inventories of radioactive materials upwind of many of our cities (Lovins and Lovins, 1982).

Fifth, this commitment has engendered financial disaster for the utility industry; the Washington Public Power Supply System default of potentially $8 billion is just the overture to what Bupp describes as "a disaster of ever-growing proportions" (quoted in Cook, 1985). There has been nearly $20 billion worth of reactors written off recently as uncompletable, unneeded or both. There is another $40 billion or so about to be written off. Something like an additional $60+ billion of excess costs must be paid to operate those plants which are now under construction and may be completed, compared with the cost of not running them and using coal power instead (although coal is too expensive to be the right competitor) (Lovins, 1985a and b).

Sixth, the federal commitment to nuclear power has yielded reactors which in general cost less to write off once they are built than to operate, and which collectively are now delivering to the country about half as much energy as wood (Lovins, 1985a and b). Altogether, the damage bill comes to over $200 billion; and the bills still to be paid for cleaning up the mess - decommissioning, waste management, uranium mill tailings, and so forth - will be of similar magnitude, assuming we can find technically and politically acceptable ways to do those tasks at all.

One can fairly describe the state of the nuclear industry as the greatest collapse of any enterprise in industrial history. The nuclear industry has indeed largely succeeded in bankrupting its only possible customer, namely the electric utilities. Rarely has so much been spent by so many for so little.

I would like to offer a personal view of the structural and conceptual errors which have led to this stupendous national potlatch - and which, if not perceived and corrected, may well lead to repetitions in agriculture, security, and other policy arenas.

Historic Roots

Nuclear energy was born with its military face uppermost, and that has influenced the development of the technology in subtle ways, not least of which is the psychology of guilt. Some people who know a lot more about the institutional history than I do have remarked that we are unlikely ever to get a sensible nuclear energy policy until everybody who was involved in the Manhattan Project has died or retired. But the flip side of this is that the Manhattan Project brought together an unprecedented concentration of technical genius which was able to do remarkable things. That very achievement led, I think, to a sense of hubris which arose from a quite justified admiration for those technical accomplishments. To this day, I think that one of the most common errors made by nuclear power advocates is to assume that there are more people of high quality available in the nuclear business than there actually are. If you spend your life at a place like Oak Ridge National Laboratory, it is natural to assume that good technical people are equally available in the Munchkinville Power Company, and that there are centers of excellence of at least the caliber of Oak Ridge which can be built in many places and which can continue to operate at that technical level indefinitely. Moreover, both in its self-enforcing assumptions and otherwise, a coherent group of people often behaves in almost tribal ways in which its members, free from peer pressure, would not dream of behaving individually.

Preoccupation with what this extraordinary group of people had been able to do in the Manhattan Project has led to the systematic exclusion of other options. In fact, as recently as 1972 the Atomic Energy Commission (AEC) was arguing that it would be wholly premature to commit $15 million to solar research - an amount far less than, for example, the Paley Commission had recommended to President Truman 20 years earlier - on the grounds that the AEC and other executive agencies considered solar energy speculative and too far in the future (JCAE, 1972a: 1141). It was necessary to concentrate on options which were proven and available. Of course, that leads to the obvious "catch 22" that the only options which are then available are the ones on which you spent all the money and expertise. (Yet ironically, since 1979, actual net orders have totaled *minus* 65,000 MWe for coal and nuclear plants but *plus* over 45,000 MWe for renewables and for fossil-fueled, decentralized cogeneration.)

Bupp and Derian (1978) have explained the history of nuclear development in the following way. Originally, the private sector was not interested in building reactors. There were several stages of increasing effort by the AEC to force private industry to build reactors, and those efforts eventually succeeded. As far as I know, no nuclear vendor in the world has made money selling reactors - although some of them have managed to turn a profit on their nuclear business by making more on repairs and fueling than they lost on construction. It was not, of course, supposed to turn out that way. Bupp and Derian describe particularly well the process of "mutual intoxication" by which the AEC, the vendors, and the utilities persuaded themselves and each other that this was just another way to boil water: it would be very straightforward, and it would be very cheap.

It would generate electricity more cheaply than coal plants, which in those days were far cheaper than they are now. The two main vendors, General Electric and Westinghouse, were essentially both cowed into entering the marketplace by the threat of a nationalized nuclear TVA, and tempted into it by lavish subsidies. They built some plants as loss leaders, allegedly to lay the foundation for a gigantic nuclear power market: Westinghouse alone reportedly lost over a billion dollars. Unexpectedly and persistently, loss leaders have since become the norm for this "market."

Looking back at the history of that period of "mutual intoxication" which led to the "great bandwagon market" for reactors, only Philip Sporn, among all the senior people in the utility industry, was skeptical enough to challenge the economic assumptions of nuclear advocacy (Bupp and Derian, 1978). All of his colleagues, in what had been a very conservative industry based on real engineering experience, simply took leave of their senses in a way that is really hard to understand in retrospect.

What has happened since that time is, if anything, worse. Figure 1 relates forecasts of installed nuclear capacity, made by various government agencies for the year 2000, to the year when the projection was made. From 1973 to 1980, there was roughly an eight-fold collapse of nuclear expectations. By now it is upwards of ten-fold; and interestingly enough, as Figures 2 and 3 illustrate, the pattern of collapse is identical throughout the world's market economies, over an enormous range of political and regulatory conditions - even in Canada, with no regulatory obstacles to building reactors, and in West Germany, with unregulated utilities. Part of the institutional setting in which it was possible to forecast in the early 1970s - just 13 years ago - that there would be about 1,200 to 1,500 large reactor-equivalents operating by the year 2000 was, of course, the AEC's charter, which established the agency to regulate *and* promote nuclear energy for both military and civilian purposes.

The AEC fissioned in 1974 into the Nuclear Regulatory Commission (NRC) and the Energy Research and Development Administration (ERDA). ERDA then fused with a number of bits and pieces of other agencies in 1977 to create the Department of Energy (DOE). The old AEC was then reconstituted in all but name in 1981. The cast of characters remained much the same throughout and, therefore, so did the perceptions which drove the federal institutions. As a result, an *electronuclear* elite has dominated U.S. energy policy for several decades - and still does. In fact, it is stronger now than it has been

Figure 1

U.S. Nuclear Capacity Projected for the Year 2000 by AEC, ERDA, and DOE

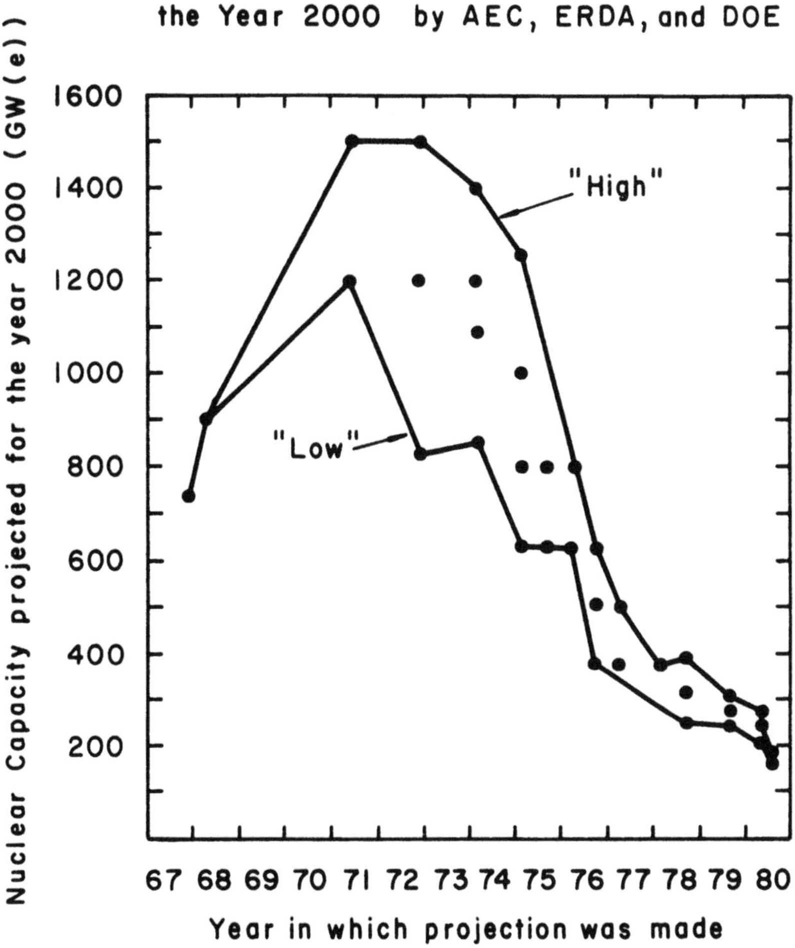

Sources: Energy Information Administration (EIA) forecasts ; and C.F. Zimmerman and R.O. Pohl, 1977, Energy 2:465-471.

Note : Forecasts after 1978 are from the EIA supply model which calculates the maximum capacity physically achievable if constraints of demand, plant siting and construction finance are ignored. Because the pre-1978 forecasts by Zimmerman and Pohl are based on the number of likely plant orders (as opposed to the physically achievable maximum), the 1978-80 forecasts exaggerate nuclear potential relative to earlier forecasts.

Figure 2

Official Canadian Nuclear Power Projections for the year 2000

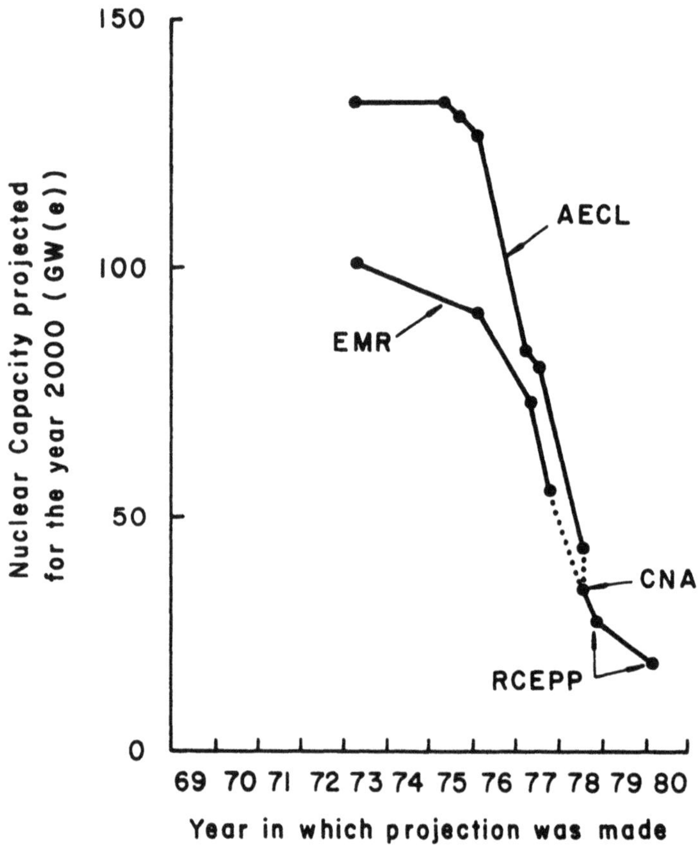

Sources : Atomic Energy of Canada, Ltd (AECL); Department of Energy, Mines and Resources, Canada (EMR); Canadian Nuclear Association (CNA); and the Royal Commission on Electric Power Planning (RCEPP).

Figure 3
Projected Federal Republic of Germany
Nuclear Power Capacity in the Year 2000

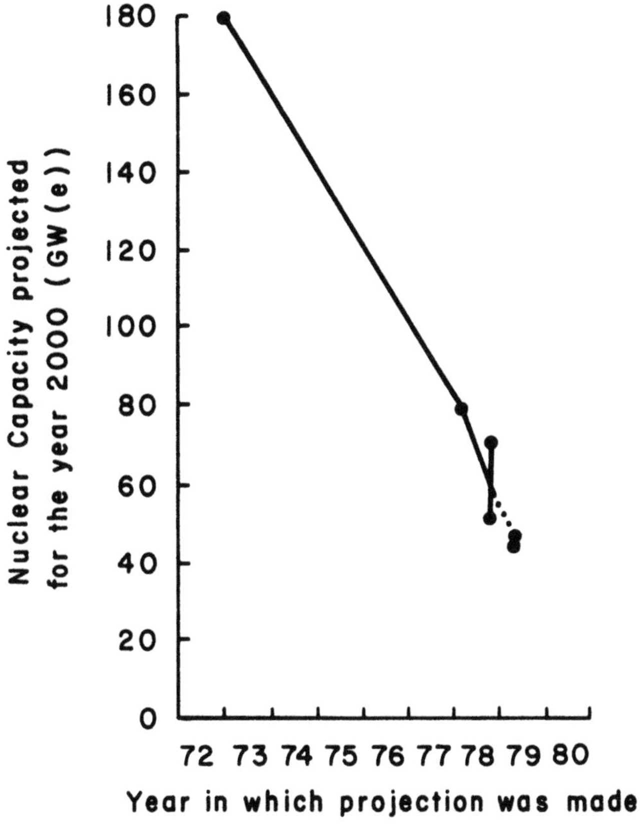

Sources: Through 1978, official forecasts of the Federal Republic of Germany. The 1979 forecast was prepared by U.S. experts participating in the International Nuclear Fuel Cycle Evaluation (INFCE).

for a decade, even though it is powerless to reverse the economic trends I will describe later, which leave this country with fewer reactors committed now than it had in 1972, and the world with the fewest nuclear starts in 1984 since 1968.

Another important institutional feature of the 1960s and 1970s which carried over into the modern Department of Energy is the preoccupation with the mission of making bombs for the military. The reason bomb-making is in DOE is that the civilian government does not want to put bomb-making in the Department of Defense, but no one is quite sure where else to put it. Historically, of course, the relationship grew out of the practice of military facilities' being owned and operated by the Atomic Energy Commission. The reasons for maintaining this practice now have more to do with public policy than with administrative convenience. The preoccupation with bomb-making - which accounts directly for 60 percent of DOE's FY 1985 appropriation - has created a deep structural flaw in DOE. It means essentially that any Secretary of Energy must be acceptable not only to the lawmakers, but also to the military. This almost guarantees that the person assuming office will reflect a number of mindsets which make it very difficult for the department to develop and follow sensible civilian energy policies.

The source of the problem is not only an insular and elite bureaucracy. The Atomic Energy Commission was pushed hard to do many of the foolish things it did by the Joint Committee on Atomic Energy in the Congress, a committee of extraordinary power and cohesiveness. The Joint Committee was symbiotic, indeed the Congress still is in many ways symbiotic, with what is arguably one of the two or three most powerful lobbies that exist - the utility-industry lobby. By some counts, there are several full-time utility lobbyists per Congressional member; and, of course, the industry's lobbying base and constituency run back into everybody's district. The utility industry has annual revenues of over an eighth of a trillion dollars. Combine this economic power with the tradition of secrecy in the military and the civilian nuclear business, and the arrogance developed by people who exercised power for many years without being accountable, and you have an excellent vehicle for making mistakes on a truly grand scale.

The promotion of myths became a central function of the bureaucracies that were themselves nurtured by the myths. Let me give four examples of how that process works.

First, there are the industry comparisons which purport to show that it is cheaper to generate nuclear than coal-fired electricity. Most of those numbers - which absolutely pervade the literature and are quoted by everybody, including the federal government - trace back to a series of annual surveys done by the Atomic Industrial Forum (AIF). A few years ago, Charles Komanoff, an independent economic consultant in New York, dug into the data base and found it was enormously skewed. The AIF choice of plants for that allegedly objective survey left out all the worst nuclear plants and nearly all of the cheapest coal plants. In fact, it left out more coal plants than it counted. There were a number of arithmetic and methodological errors as well. Komanoff showed that when all plants are counted on both sides, nuclear plants have sent out costlier electricity than contemporaneous coal plants in each year since about 1975, and that this margin has been rapidly increasing (Komanoff, 1980).

A second example concerns capacity factors, which are very important for evaluating the economic viability of something as capital-intensive as a power plant. It had been standard procedure to assume that nuclear plants would send out 80 percent of their full-time/full-power output. Some utilities still assume 75 percent. The empirical number for all large reactors in the U.S. to date is a little under 55 percent and is probably falling. When it was pointed out by the late David Comey (1974: 23) that the plants were not nearly as reliable as expected, the federal government simply redefined the norm in good Soviet fashion by inventing something called "maximum dependable capacity" - which is how much the plant can send out on the hottest day of the summer - and re-evaluated all the capacity factors on that basis rather than using the original design rating by which the plants had first been justified. (Similarly, the Edison Electric Institute has lately been so embarrassed by national reserve margins in the vicinity of 40 percent that it has started using instead a similar-sounding ratio which divides excess capacity - peak capability minus peak load - not by the peak load but by the peak capability, which is a larger denominator. Presto! The overcapacity sounds much more modest.)

A third case involves the well-known statistic that the coal fuel cycle, with its air pollution and other problems, is more dangerous publicly and occupationally than the nuclear fuel cycle. The American Medical Association, among others, has published a comparison of this kind. But these comparisons fail to include risks after 30 or 40 years. Of course, that omission disproportionately emphasizes risks associated with coal, where most of the damage is short-term (e.g., a prevalent health risk is emphysema, which typically appears relatively early after exposure). Conversely, nuclear power is made to appear attractive because nearly all of the officially calculated risk is from radon from mine tailings, carbon-14 from reprocessing, and other long-term hazards. Ramsay and Russell have examined the official surveys that are considered authoritative by proponents of nuclear power, such as the Rasmussen Report on reactor safety and the EPA fuel cycle analyses (1978: 387-403). They assumed that those studies were right, ignored climatic and proliferation risks as unquantifiable, and calculated the total public and occupational risk of the coal and nuclear fuel cycles as a function of discount rate. If future deaths are discounted at 5 percent per year, then risk is effectively ignored after a few decades and nuclear appears to be about ten times as safe as coal. If, however, a zero discount rate is employed, as is theoretically and morally required - that is, weighting long-term harm as being as important as short-term harm - then the result reverses by *three orders of magnitude.*

Finally, there is the notion encouraged by the utility industry and the federal government that the high burn-up plutonium made in power reactors is incapable of being used for powerful or reliable bombs. This has been known in the weapons community to be false since at least 1962, when a Los Alamos test settled the matter empirically. Its falsity was widely noted in the open literature by leading bomb designers during the 1970s; yet it remained the basis for U.S. nonproliferation policy throughout the period. In a review article in *Nature,* I sought to expose this myth and dispose of the matter; yet the same myth persists to this day in some rather high circles (Lovins, 1980b). The upshot of this kind of myth-making is that the nuclear establishment has succeeded in concealing the truth even from itself, and hence has spread "civilian" bomb kits around the world in the erroneous belief that bombs could not hatch out.

A side-effect of such self-delusion, incidentally, is that the Reagan administration - the most outspokenly pro-nuclear in American history - probably believes sincerely that it is helping its favorite technology. My own perception is that President Reagan and his Department of Energy have contributed signally to the irreversible meltdown of the industry's public credibility: as Mark Twain remarked, "A cat which sits on a hot stove lid will not do so again, but neither will it sit on a cold one." In one of the more bizarre instances, the industry orchestrated the President's remark that nuclear power was "hamstrung" by inept regulation, and instantly splashed the quotation across the nation in preset full-page ads. Commissioner Gilinsky of the NRC, however, pointed out to the American Nuclear Society that such an attack on the regulators did not seem to him very smart: the public, he said, may not trust the NRC very much, but they don't trust the industry at all, and if they come to believe that the NRC is incompetent, they'll conclude, not that nuclear power should be regulated less, but that it's even more dangerous than they thought. He was shortly proved right.

Economic Myths

Nowhere are the consequences of nuclear myth-making more clearly evident than in the economics of nuclear power. Although the future cost of nuclear electricity depends on some fourteen major and thirty-odd minor variables, all of which are disputed and many of which are highly uncertain, capital cost is the most important single variable. There has been a series of very detailed studies over many years by the AEC and its successors, and by major nuclear contractors, of how much it would cost to build a reactor in constant dollars per installed kilowatt (EIA, 1982: 34). Those studies have all been wrong, quite often by factors of 3 to 8 - very large numbers. Komanoff again happens to have assembled the empirical data and painstakingly discounted and deflated everything to constant 1979 steam-plant dollars per installed kilowatt (Komanoff, 1981). He finds a supply curve looking like that given in Figure 4. The solid portions of the curve represent all of the nuclear and coal plants commissioned between 1971 and 1978 in the United States. The dotted curves represent an extrapolation of those trends, if moderated. The crosshatched areas represent the Department of Energy's official assumptions about what will happen. The "X" represents what was actually happening in the nuclear sector as of mid-1983, in contrast to coal where the data have tended to track the dashed curve quite closely. Notice what is happening here. The real cost per kilowatt for both kinds of plants is increasing with the number of plants built or being built. This is a classic economic supply curve. It is upwardly monotonic. That was not supposed to happen. The real cost was supposed to be constant or indeed to decline as a result of supposed learning effects.

What seems to be happening can be summarized by a hypothesis which was developed by Bupp et al. (1975) and later refined by Mooz (1978) and Komanoff (1981) in collaboration with Vince Taylor. Normally if people think an activity is hazardous, the market signals that perception through things like insurance premia and tort liability. In the case of coal plants, the signaling is done at least in part through the Clean Air Act (an unintended side-effect of which, incidentally, is acid rain). In the nuclear case, however, the signals are largely suppressed by the Price-Anderson Act, which sets an absolute ceiling on liability for nuclear accidents.

Figure 4
Empirical Supply Curves for
U. S. Central Power Stations

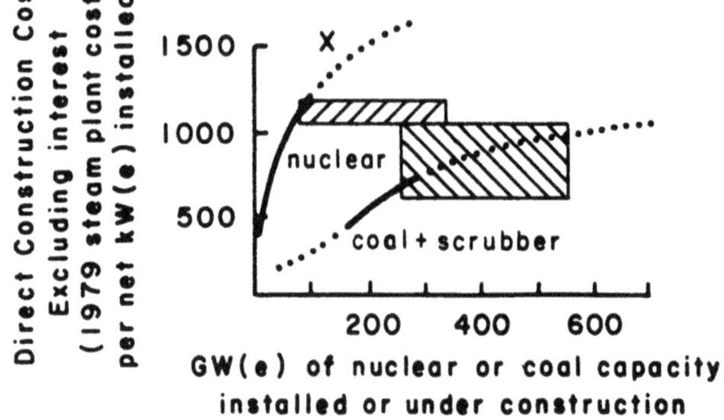

——— Actual costs for plants commissioned during 1971-78. Real capital costs per kW rose 142% (13.5% per year) for nuclear plants and 68% (7.7% per year) for coal plants. Both figures refer to construction costs prior to the Three Mile Island accident.

.... Projection from Komanoff's multiple regressions. Nuclear projection is based on data for all nuclear plants during 1971-78 (R^2 = 0.93). Coal projection is derived from 116 comparable coal plants as defined by Komanoff (R^2 = 0.68).

X Average mid-1983 completion cost estimate by utilities engaged in construction of nuclear plants.

[hatched box] U.S. Department of Energy Electricity Policy Project assumptions (1983).

Sources: Charles Komanoff, 1981, Power Plant Cost Escalation: Nuclear and Coal Capital Costs, Regulations and Economics; and DOE, 1983, Future of Electric Power in America: Economic Supply for Economic Growth (June) DOE/PE-0045.

Note: For both types of plants, the figure indicates that the cost per kW increases in real terms as more plants are ordered.

Now, the more plants are built, whether coal or nuclear, the more air pollution (or other perceived hazard) is created, and the more political pressure will be put on the regulatory process to make each plant cleaner or safer in order that the perceived hazard from the larger total population of plants does not increase. Also, the more plants that are built, the more likely it is something will go wrong with one of them and that it will be near populations when it does, causing the communities to take notice. This intensifies popular interest in the subject and, one would expect, increases regulatory pressure to tighten safety or air-pollution standards. Meanwhile, returns on plant investment in cleanliness or safety tend to diminish. You would therefore expect the real cost of each plant to rise geometrically with the numbers of plants built. That is precisely what we observe. In fact, we observe it with a goodness of fit of 93 percent for the nuclear plants and 68 percent for the coal plants. This is a much better fit than any other explanation can command. This does not mean that the explanation is right - correlation never proves causality - but what it does suggest is there is a burden of proof on people who have some other interpretation to show that their alternative fits the data at least as well.

If the Bupp-Derian-Komanoff-Taylor interpretation is correct, it is not very good news for nuclear power. Reducing the disparity in slope between the coal and nuclear supply curves (nuclear being much steeper) appears unlikely. Why? Because in the case of coal, the perceived hazard we are trying to abate (with investment in scrubbers and so on) is very tangible: one can see it, smell it, and wipe it off the windowsill. But in the nuclear case, the hazard is ineffably abstract. While engineers may succeed in reducing the calculated risk of a meltdown from 10^{-x} to 10^{-y} per reactor-year, popular concern is unlikely to be allayed; such alleged reductions in a supposedly unlikely but catastrophic risk have no political meaning. People tend to be much more interested in the large consequence than in the small probability and, therefore, cannot be expected to lower their concerns. Thus, society in the future will likely require bounded additional investments to abate coal hazards but open-ended ones to abate nuclear hazards. It is intriguing, too, that if supply curves are prepared using data from the past two years for such disparate countries as Canada, Britain, France, Germany and the Soviet Union, they will be strikingly similar to the U.S. ones (Harding, 1985). In all of those countries, even the centrally planned economies like France and the Soviet Union, there is now very rapid real cost escalation, apparently for quite similar reasons.

Another way in which we have managed to conceal economic reality from ourselves is through generous subsidies, which are particularly heavy for nuclear plant construction. Roughly 42 percent of the marginal cost of nuclear construction is socialized through tax preferences and another 28 percent or so is deferred to future ratepayers - a total distortion of approximately 70 percent. Thus, while the average delivered marginal price is about 20 cents per delivered kilowatt-hour in 1984 dollars, power can be sold at an apparent price of 6 cents/kWh. If it *looks* that cheap (if we do not know what it really costs) then of course we do not know how much is enough, and we underinvest in energy productivity. Subsidies have led utilities down the path to ruin by encouraging and enabling them to build more power plants - perhaps $100-200 billion worth more - than they will be able to amortize from revenues. The Reagan administration has substantially *increased* such subsidies by every means from illegal enrichment discounts to 10-year nuclear depreciation, Shoreham and TMI bailout legislation, utility dividend reinvestment deductions, and Internal Revenue Service (IRS) permission to finance many nuclear construction costs with tax-exempt pollution-control bonds (because the radioactivity would be a pollutant if released).

The technological history of U.S. nuclear power, like its economic history, suggests a systematic effort to avoid reality. There are a great many conceivable types of reactors: Theodore Taylor has estimated at least on the order of 10^{12} (1984). This number is probably larger when all the possible combination of fuels, moderators, neutron spectra, and coolants are considered. The choice in this country, however, was narrowed very rapidly, mainly under the exigencies of military design. Water cooling began with a decision to locate plutonium production reactors on the Columbia River. U.S. reliance on light-water reactors is largely an outgrowth of decisions concerning nuclear applications to run aircraft carriers. The aircraft-carrier design was put on land and made into a power reactor. The sodium-cooled reactors evolved from their use to power early nuclear submarines. These were not quite as successful, but the technology was still tried on land in what we now call liquid-metal fast breeders.

It is remarkable how a very conservative industry accepted a wild extrapolation of plant size in just a few years: from plants that people had experience with, typically in the 100 or 200 MWe range, to plants in the 800 to 1200 MWe range, which were ordered in very large numbers before such a thing had ever existed anywhere. That is one of the reasons that the whole process came unstuck: at that scale, complexity and new problems arose which were wholly underestimated. Technologists who had previously required themselves to be able to say "We have solved the problem" or "We know how to solve the problem" became intellectually sloppier, content to settle for "We think we will find a way to solve the problem" or, worse, "We know enough to form a judgment that we have a good basis for feeling competent that we will be able to solve the problem." All too often, they did not turn out to know that much, and the cause of problems turned out to be prior solutions.

Safety regulation, in my view, has never been a significant constraint on reactor choice or deployment in this country, simply because the economic pressures on the regulators and their political composition were such that ways were practically always found to issue the licenses. Whenever a problem was discovered which applied to more than one reactor, it would be ruled out of order in the licensing proceedings for any particular reactor, or it would be labelled a "generic issue" and then put aside to be addressed some other time. The list of unresolved generic issues has grown steadily for years. The result has been that when the NRC attempts to resolve one of those problems, it becomes necessary to require retrofits of existing plants and redesign of plants already under construction. This is richly productive of delays and cost overruns. The industry's complaints about the "ratcheting" and "instability" in the regulatory process miss the point. That instability derives from premature deployment of a technology that is not yet ripe. If it were mature - if there were no surprises left - regulation could be stable.

Fuel Cycles

It is important to look at the technological imperatives that drove U.S. fuel-cycle development, because the assumption all along was that it would be essential to move from burner to breeder reactors. This, of course, meant a transition from a uranium to a plutonium economy and therefore to a closed fuel cycle which would reprocess the spent fuel and extract the plutonium for reuse. Hence the only kind of waste management to be considered would be that which dealt with the products of reprocessing - a very complex process which chops up the fuel, dissolves it in acid, chemically recovers the plutonium and unburned uranium, and then isolates the high-level waste from the medium- and low-level wastes and the solid wastes from the liquids and from the gases. This logic, which shaped U.S. strategies for fuel and waste management, unravelled during the 1970s.

First, it was discovered to be hopelessly uneconomic to recycle plutonium in ordinary thermal reactors. Second, the need for the holy grail that drove this whole business, namely, the fast breeder reactor, became subject to searching questions that could not be satisfactorily answered - and not only on grounds of safety and proliferation (Lovins, 1974). As early as 1972, it was pointed out that, according to the official numbers, the price of nuclear electricity would be very insensitive to the price of uranium, and therefore you could afford to search all over for uranium and mine it at a much higher price without making much economic difference (Lovins, 1980a: 70). Cochran also pointed out in a series of cogent studies starting in 1972 that breeders would have such a high capital cost that it would at least offset their proposed saving on fuel, so that the economic need for fast breeders would be pushed at least a century into the future, if not more (see, e.g., 1974). A series of studies by Brian Chow (1979 and 1980) nailed down this conclusion permanently.

Additional arguments were presented. It was noticed in the mid-1970s that the roughly 50-fold improvement in uranium utilization which in principle might be provided by fast breeders would take one or two centuries to achieve, because it would take that long for the fuel cycle to come to equilibrium. In round numbers, it takes about seven tons of plutonium to start up a 1,000 MWe breeder reactor, just to provide its initial core and the pipeline inventory. But running a light-water reactor produces only about a quarter of a ton of plutonium per year. Thus, thirty or so reactor-years of operating conventional reactors are needed to provide sufficient plutonium to fuel *one* breeder. Moreover, the breeders built to date have all been sterile, in the sense that they have not reproduced their initial fuel load fast enough to double it during their lifetimes. Given the thirty or so reactor-years of operation needed to produce the plutonium for one breeder, it appears that if uranium is going to be as scarce as breeder advocates were saying, then it is already too late for the breeder, because so much uranium is needed to run the thermal reactors to fuel the breeders.

During the next eighty years, the uranium saved with breeders (which is only a several-fold saving) could be achieved much more cheaply and surely with uranium-efficient thermal reactors. With present light-water reactors, 15 percent of the uranium could be saved with retrofits; 30 percent in new light-water reactors; and about 40 percent in existing CANDU reactors with low enrichment (Feiveson et al., 1979). If the government insisted upon recycling uranium, but threw away plutonium, nearly 80 percent of presently used uranium could be saved (Feiveson et al., 1979). The added duration of uranium resources would actually be much higher, because these options not only stretch finite uranium supplies, but also make nuclear power less sensitive to uranium prices. If the supply of uranium increases only linearly with price, a 40 percent improvement in the efficiency of using uranium is equivalent to nearly a trebling of the electricity manufacturable at a given uranium price per kilowatt-hour. Thus the 80 percent recovery of uranium possible with non-plutonium, non-breeder fuel cycles using presently commercial reactors can stretch uranium supplies more than twenty-fold.

If uranium scarcity were still a political concern even after reactor efficiency were improved, the long-touted storage advantage of uranium over oil could be cited in support of nuclear burner reactors. With a ten-year stockpile of low-enriched uranium and with uranium oxide (U_3O_8) at $40 per pound, carrying charges on that stockpile would increase delivered electricity prices by less than one percent with the least uranium-efficient reactors known, or by something like one-quarter or one-fifth of one percent with the most uranium-efficient reactors. The sort of uranium stockpiling which Japan has been doing for a long time would be more rational economically than breeders, *assuming* the problem is a potential disruption of uranium supply. Of course, even neglecting all of these arguments, whatever energy security one might gain some time after the year 2100 from fast breeders would only address baseload electricity, which is only about 4 percent of U.S. delivered energy needs. Such "security" would be purchased with the added energy vulnerability inherent in central-electric systems and especially in plutonium fuel cycles, where a breakdown in one reprocessing or fuel-fabrication plant could idle about forty or fifty large reactors at a stroke. These arguments were expressed cogently during the 1970s, especially by the Princeton group (Feiveson et al., 1979). Ignorance is no longer an available defense for nuclear policymakers. They did not, however, acknowledge *any* of these breeder-economics arguments; they were simply overtaken first by the wider collapse of nuclear economics and indeed of central-power-station economics generally.

Meanwhile, practicality also intervened: European experience was showing that reprocessing is a great deal more expensive and troublesome than had been thought (Patterson, 1985). The flagship French plant at Cap La Hague has processed approximately one-tenth of what its builders projected, and is having increasingly serious operational and safety problems. (Some of these are caused by insoluble granules of refractory metals, which, interestingly, would be at least an order of magnitude more plentiful in plutonium fuel than in the uranium fuel reprocessed so far.) No country in the world has made a reprocessing plant work on a commercial basis, and the contract price for reprocessing in France and Britain has increased dramatically; the French price recently increased by a factor of seventeen in nominal francs over ten years. The doubling time of the real cost increase has at times been less than a year and may be getting shorter. In fact, French and British reprocessing contracts reportedly contain a clause allowing the contractor, for any reason or none, to not reprocess, return the fuel to the customer in whatever condition it happens to be in, and keep the customer's money - hardly an indication of confidence in the technology.

Reprocessing also turns out to make waste management a great deal more difficult. It results in thirty to fifty times the initial volume of waste materials which have to be looked after for a very long time. Furthermore, reprocessing produces a diversity of wasteforms. Some of the wastes have high thermal power density; some are gases, including labile gases like tritium; some are corrosive; and some have enormous surface area (all of the old paper towels, rubber gloves, and bits of plumbing that are called "low-level transuranic" waste). Taking these factors into consideration, reprocessing appears to make the waste management problem more intractable than simply packaging spent fuel and storing it - although we have yet to determine where that might be (Patterson, 1985; Hatzfeldt, 1979).

There is no question that the waste can be stored safely in the ground; the concern has always been how long it will stay there. Yet the Department of Energy has returned to the position held by the AEC in the early 1970s, asserting without a storage plan and site in place that the waste problem is solved. DOE's "policy" is that we will reprocess, as if anybody were there to do it. Supposedly, the high-level waste will be converted into borosilicate glass and buried in salt. Yet it has been known for seven or eight years that borosilicate glass and salt is the *least* compatible known combination of wasteform and host rock. When the glass is put in salt, the water in the salt migrates toward the heat and makes brine. Hot brine rapidly dissolves virtually anything (except gold). In fact, it decomposes borosilicate glass in a matter of weeks.

We have come full circle to where we were over a decade ago in understanding the waste management problem, yet we still have essentially no respectable technical base for it that does not run through the reprocessing route. All that we have to show for the effort is a scientifically designed white elephant. The breeder reactor accumulated the largest R&D expenditure on a single project in U.S. energy history, but its proponents failed to address the right problems.

Technology Is The Answer! (But What Was The Question?)

With the demise of the Clinch River project in 1984, we now see more clearly the rapid emergence of nuclear fusion as the second-biggest money-gobbler in the federal civilian energy budget ($440 million in the FY 1985 appropriation). The fusion program has become in part a welfare program for unemployed fission technologists. If they are successful in developing the technology (which will be much tougher than putting people on the moon) and if waste, safety and safeguards problems can be made tractable (which is by no means assured), fusion will then be an even more clever way than fission to do something we do not need to do - manufacturing more big blocks of expensive electricity for which there will not be a market. For that reason, fusion will not be economically sound and, therefore, little point exists in arguing about its safety, environmental value, or beauty. We just should not support the program with public money.

The shift in emphasis from breeder to fusion technology - along with some rather expensive wishful thinking DOE is encouraging about "next-generation" fission reactors, as if there were going to be any - recalls Santayana's remark that "fanaticism consists in redoubling your efforts when you have forgotten your aim." I think that is exactly what the technological history shows in the case of nuclear power: repeated efforts to invent a better horse and buggy. Yet technological history is only the most superficial layer of what went wrong. More profoundly, the federal energy establishment fundamentally misunderstood its mission. It perceived (and still perceives) its mission as the development of a new piece of hardware, a kind of magic bullet, in which the economics would take care of themselves. The rationale seems to be that as long as the technology is sufficiently sweet (meaning that it delivers large chunks of electricity from something ingenious and complicated which employs many high technologists for decades or, preferably, millenia), the hardware magically would be delivered, used, and preferred. This manner of thinking fundamentally mistakes what the energy problem is, and what the federal role should be in addressing it.

American society and its economy do not require electricity for its own sake. Raw kilowatt-hours are useful to nobody but executioners. There are uses of energy for which electricity is the only or the best form, and there is no question that electricity is a very high-quality form of energy. Our problem, however, is not a shortage of electrical generating capacity. If all power plants already built in the U.S. could sell all of their output when operated at capacity, approximately 16 percent of delivered energy in the U.S. would then be in the form of electricity. Currently, only 13 percent of delivered energy is electrical, the other 3 percent representing idle generating capacity. Electricity is very expensive, with the national average price for all sectors nearly 7 cents per kilowatt-hour - the equivalent, in terms of heat content, to oil at about $120 per barrel. Therefore, the uses for which electricity is economically worthwhile are limited to special ones like motors, lights, smelters, and so on. Electricity is fundamentally uneconomical in thermal markets (58 percent of our delivered energy needs) and in road vehicles (about 34 percent of our needs). Only about 8 percent of our delivered energy needs can be economically met by electricity even at present (very low) levels of end-use efficiency. Thus, the U.S. *already* has roughly twice the electrical generating capacity that it needs.

In this respect, the market for electricity is and has been oversupplied. Yet, electricity has historically taken on the order of three-quarters of the energy R&D budget - never less than 60 percent. In the 1984 budget request from DOE, for example, the civilian energy budget (of which roughly one-third was for basic research and two-thirds was for hardware) dedicated over half to nuclear-related electricity projects alone. Of the entire civilian energy budget, 65 percent was nuclear - not counting some things on the nuclear side that are military (like laser fusion). Of all the hardware research, 92 percent was for electricity. Of all the energy supply dollars, 96 percent was for electricity (DOE, 1983a). Of all the electric dollars, over 99 percent was nuclear. All of that funding for 8 percent of the problem!

Electricity will not be made more competitive by increasing plant generating capacity. But just such an assumption is embodied in federal electricity R&D budgets. There are more than 50 kinds of diseconomies of scale which have to be taken into account along with the several economies of scale, because what we are interested in is *net* economics, not gross (Lovins and Lovins, 1982: Appendix A). It has already been carefully shown that - except in very unusual cases like running a smelter - it is much cheaper to build small power plants than large ones (Lovins and Lovins, 1982: Appendix A). It would be also far less risky for the utility. The Edison Electric Institute now admits this reality and favors only small, fast, cheap investments (Bauer, 1984).

Will we need more electricity in the next decades? There are light bulbs now available with four times the efficiency of the standard bulb. These bulbs render obsolete about 25,000 MWe of power plants - $60 billion worth at today's marginal prices. There also are high-frequency ballasts, paying back their cost in about a year, that can be installed in overhead fluorescent lights. Those render obsolete about another 60,000 MWe. Energy-efficient refrigerators remove the need for 20,000 to 25,000 MWe. Alone, these refrigerators displace roughly one-half of the nuclear capacity today. In fact, *all* U.S. nuclear capacity could be replaced just with proper sizing, coupling and controlling of industrial electric motors, saving about 70,000 MWe installed at a cost of under one cent per kilowatt-hour (Lovins, 1985a and b).

If full use were made of the best electricity-saving devices now available, that would at least quadruple the efficiency of using electricity in this country, at a cost of less than two cents per kilowatt hour in 1984 dollars (Lovins, 1985a and b). That is less than just the *running* cost of any thermal power plant, including a nuclear one. As a consequence, once Seabrook I or Palo Verde or Diablo (or any other nuclear plant in the construction pipeline) is completed, it will be cheaper to write it off and buy efficiency than to operate it (Lovins, 1984 and 1985a and b). Between 80 and 90 percent of the electricity now sold in this country is or will shortly be uncompetitive with the efficiency improvements now on the market. Nonetheless, we continue to devote nearly all of the federal R&D budget to developing more ways to make expensive electricity - more ways to make the unsellable. In the case of power plant R&D, it means the development of the unfinancible to make the unsellable.

The Department of Energy persists in forecasting high electricity demand - so high, indeed, that about two-thirds of all primary energy growth to the year 2000 is officially projected to go to conversion and distribution losses (mainly cooling-tower plumes and warm condenser water) and never to get to the consumer at all! Indeed, a 1983 DOE report argued that to keep the lights on in the 1990s will require utilities to order another trillion 1982 dollars' worth of power plants over the next couple of decades (DOE, 1983b). That amounts to ordering big power plants at a rate averaging one a week, in a country that has ordered only a handful of smallish coal plants since late 1981. The Fourth National Energy Policy Plan declared that it saw little or no workable competition for electricity (discussed in Lovins and Lovins,

1984). That is not much further on than we were in 1972, when the AEC was unable to cite a single literature reference on the possibility of saving energy (JCAE, 1972b: E1109). A detailed government report called *A New Prosperity,* produced by the Solar Energy Research Institute in 1981, is apparently on DOE's index of prohibited reading (see discussion in Lovins and Lovins, 1984). It is not even mentioned in DOE's forecasting documents, even though it is the best end-use/least-cost analysis that has been done in this country. The evangelical theology which holds that electric use and GNP must spiral ever upwards in a frenetic embrace continues to dominate our energy policy. It comes in part out of the Office of Policy, Planning and Analysis at DOE, a name which I think correctly reflects its priorities: policy first, analysis afterwards.

The Department of Energy and its predecessors have assumed that all that is needed is to develop the right kind of power plant, and the utilities will buy it and get the electricity out there. Having failed to resist the department's strategy, the average utility now is bankrupt in all but name. It spent the 1970s increasing its construction expenditures eight times as fast as its cash earnings, and borrowing about two-fifths of its dividend payments. In effect, this industry has gone broke, liquidating itself to build power plants that it does not need, cannot afford, and will not be able to pay for. It has been playing "You Bet Your Company" that it will take customers 50 years to discover better buys. It is not taking customers nearly that long.

Utilities are competing in a highly competitive energy-service marketplace. They have to compete with weatherstripping, with motor controls, with better ballasts and lights. If they fail to do so, they will simply find (as they already have begun to) that higher prices *reduce* their long-run revenues. Their product is already so uncompetitive that pricing it higher loses market share.

There are some very exciting things utilities can do (as a few are doing) to participate financially in efficiency improvements so that the industry again can become a declining-cost industry and so that firms once more have reasonable cashflows (Lovins, 1985a and b). The Department of Energy is not in the least interested in that sort of survival strategy. But the economics which DOE has long ignored actually are paramount. Since 1979, this country has gotten *over one hundred times* as much new energy from savings as from all net expansions of supply combined (Lovins, 1985a and b). More new energy has been developed from renewable sources than all of the non-renewables. More new generating capacity has been ordered from small hydro and windpower than from coal or nuclear plants or both, without even counting cancellations (Lovins, 1985a and b).

If you want to gauge where the market is headed, consider California, where since 1981 the utilities have been offering to buy privately generated, privately financed power for a little less than it would cost them to make it themselves. At the end of 1984, some 20,300 MWe of privately generated power had been firmly offered, over half of it from renewables (Lovins, 1985a and b). The principal supply sources were industrial cogeneration (some of it biofueled), wind (4,000-plus MWe), small hydro, and geothermal. But there is every other source in there also, from photovoltaics to generating energy from burning peach pits (or energy studies). The proffered supply at less than full avoided cost has lately been increasing at the rate of nine huge plant-equivalents per year. It is coming in so fast the California utilities do not know what to do with it; and the sudden glut, with enough small-power production already on offer to displace three-fourths of the state's entire peak output from nonrenewable sources, forced the suspension of small-power contracting in April 1985. Yet, two dozen states and provinces still hope to sell California *their* surplus power, all simultaneously.

Nonetheless, DOE steadfastly ignores economic reality, continuing to subsidize nuclear power heavily. Preliminary results of a comprehensive analysis of federal energy subsidies, being conducted by Heede (1985) at Rocky Mountain Institute, suggest how misguided national policy has been. Heede estimates the subsidy *just in* FY 1984 (major programs only) to be *$46 billion,* of which civilian nuclear fission received $15.8 billion. In terms of million Btu supplied per 1984 dollar of subsidy, nuclear power subsidies are calculated to have "produced" 0.07 per dollar, compared to 0.4 for hydroelectric, 1.7 for other renewables, 2.4 for oil, 3.8 for natural gas, 5.8 for coal and 13 to 26 for end-use efficiency. Thus in 1984 a dollar spent on subsidies to efficiency and renewables yielded about 80 times as much energy as a dollar spent on nuclear power. Per unit of energy, counting electricity at its heat value, electricity was 11 times as heavily subsidized as direct fuels, and at least 48 times as heavily subsidized as end-use efficiency.

Mistaken Mission

The fatal error made by DOE regarding nuclear power was supposing that the problem was hardward, not social. The public has perversely refused to evaluate and accept nuclear power as the department's analysts thought they would (or should). This has been variously ascribed to ignorance or hysteria (and indeed DOE hired a psychiatrist in 1984 to access the public's irrational "nuclear phobia"). We have also found *within* the nuclear business that the paramount problems are those of people, not of engineering. In fact, some of us are particularly concerned about the declining quality of the people going into that business: it is passing, as Nobel physicist Hannes Alfvén told the 1980 Geneva IAEA conference, "into ever less competent hands." Probably the best half of the people in the industry have already departed. It will become even harder in the future to get talented people to devote their careers to looking after the garbage; it is just not a very interesting thing to do. A dollar invested now in solving the unsolved problems will buy less solution than it did in the days of the pioneers.

DOE has never compared hardware options symmetrically with policy actions - for example, accelerated scrappage of cars or better information programs on efficient appliances versus developing some new kind of gadget with brass knobs all over it. Nor has the Department ever symmetrically compared supply expansions with energy-productivity improvements, least of all in cost.

It is difficult to find a major public-policy decision about energy in this country that was based on economics. The decisions have instead been based on political expedience, and then the economics were made to come out to justify what was done (usually by juggling subsidies). In that mode, federal energy policy is increasingly irrelevant, and will continue to be so until it addresses what is happening in the marketplace and why. The Department of Energy and its predecessors have never put down on one piece of paper the marginal cost of providing a given energy service from a variety of measures so that it is possible to compare them. This is absurd and a cause for national shame.

The Department of Energy has never considered the political impacts of its policies, although those are no less real than any other kinds of impacts. Neglecting them has resulted in many DOE-developed technologies' coming unstuck. A whole constellation of concerns has been ignored. How are the risks and impacts of the technology perceived? Are they relatively understandable? directly sensible? Are they susceptible to commonsense judgments by generalists? Is the proposed technology or action potentially vernacular, accessible to many actors (and many levels of sophistication or complexity), or is it just the exclusive realm of a technological priesthood? In political terms, is the technology centrist or centrifugal, autarchic or localist, implemented by fiat or by market choice, technocratic or accountable, loose-fitting or unforgiving, comprehensible or arcane? How does it affect federal versus state or local roles, settlement patterns, big versus small business, homogenization versus pluralism, rural versus urban, distributional equity, interregional and intergenerational equity? Does it allocate the social costs of energy essentially to the same people at the same time, or to different groups at different places and times? These are really key questions for policy. They've never been *asked* in DOE, let alone answered.

In short, I think the Department of Energy and its predecessors came unstuck by never asking the basic questions:

o What do we want the energy for?

o What is the amount, type and source of energy that will do each task in the cheapest way - cheapest in the broadest sense?

o What actions are required? Are they hardware or policy actions?

Once you have answered these questions, *then* you can start to figure out what the federal government needs to do. In fact, the Department of Energy has taken very nearly the opposite approach. DOE *starts* with existing or hoped-for bits of R&D which reflect contractors' and technologists' "wish lists" - what they would most like to play with. Then DOE tries to find a use for those bits of hardware. And, lastly, belatedly, often unsuccessfully, it tries to figure out how to make the marketplace absorb the hardware. As Herman Daly once remarked, that procedure "is unworthy of any creature with a central nervous system, let alone a cerebral cortex. And for those of us who have souls," he added, "it is incomprehensible in its inversion of means and ends."

This is the first time in many years that I have written on nuclear issues (other than proliferation). I try to avoid the subject for three reasons. First, it is so dead that its only value is as a cadaver for policy autopsy, and I would not want to be thought so dull as to be spending a lot of time kicking a dead brontosaurus. Second, it always saddens me to think of the lives, careers, money, and other resources that this national tragedy has consumed. Third, thinking about that tragedy makes me feel that there but for the grace of God go I. For in my student days I received national awards for nuclear-physics research from, among others, Westinghouse, General Electric, the AEC (Glenn Seaborg was the judge), and the American Nuclear

Society. My local utility sent me to the National Youth Conference on the Atom. When all I knew about nuclear power was what its promoters told me - which I uncritically accepted until nearly 1970 - I thought it sounded like a great idea. I doubt those promoters knew better themselves: they simply repeated what their peers all said and what they so desperately wanted to be true.

In that spirit of sad respect for the ordinary human frailities which produced the nuclear fiasco with such Sophocleic inevitability, I would like to close with an epitaph which I wrote some years ago for the nuclear industry:

> When the history of the nuclear controversy comes to be written, those who killed nuclear technology will be seen to have been its most avid promoters, who systematically mistook hopes for facts, advocacy for analysis, commercial zeal for national interest, expertise for infallibility, engineering for politics, public relations for truth, and the people for fools.

REFERENCES

Battelle Pacific Northwest Laboratories, 1978. *An Analysis of Federal Incentives Used to Stimulate Energy Production.* PNL 2410-REV.

Bauer, Douglas. 1984. Remarks at "The Energy Utility Industry in a Time of Change" seminar. Brookings National Issues Forum, April 25, at the Brookings Institution, Washington, D.C.

Bupp, Irvin C. and Jean-Claude Derian. 1978. *Light Water: How the Nuclear Dream Dissolved.* New York, NY: Basic Books.

Bupp, Irvin C., Jean-Claude Derian, Marie-Paule Donsimoni, and Robert Treitel. 1975. "The Economics of Nuclear Power." *Technology Review* 77 (4) (February): 15-25.

Chow, Brian G. 1980. "Comparative Economics of the Breeder and Light Water Reactors." *Energy Policy* 8: 293-307.

_______________. 1979. "Economic Comparison of Breeders and Light Water Reactors." Report AC8NC113 to the U.S. Arms Control and Disarmament Agency. Marina del Ray, CA: Pan Heuristics.

Cochran, T.B. 1974. *The Liquid Metal Fast Breeder Reactor: An Environmental and Economic Critique.* Baltimore, MD: Johns Hopkins University Press for Resources for the Future.

Comey, David Dinsmore. 1974. "Will Idle Capacity Kill Nuclear Power?" *Bulletin of the Atomic Scientists* 30 (9) (November): 23-28.

Cook, James. 1985. "Nuclear Follies." *Forbes* 135 (3) (February 11).

DOE (U.S. Department of Energy). 1983a. Energy Research Advisory Board. *The Federal Role in Energy Research and Development* (February) DOE/S-0016. Washington, DC: U.S. Government Printing Office.

______________. 1983b. Office of Policy, Planning and Analysis. *Future of Electric Power in America: Economic Supply for Economic Growth. Report of the Electricity Policy Project* (June) DOE/PE-0045. Washington, DC: U.S. Government Printing Office.

EIA (Energy Information Administration, U.S. Department of Energy). 1982. *Projected Costs of Electricity from Nuclear and Coal-Fired Power Plants, Volume 2* (November) DOE/EIA-0356/2. Washington, DC: U.S. Government Printing Office.

Feiveson, Harold A., Frank von Hippel, and Robert H. Williams. 1979. "Fission Power: An Evolutionary Strategy." *Science* 203 (4378) (January 26): 330-337.

Harding, James. 1985. *Awakening from the Nuclear Dream.* San Francisco, CA: Friends of the Earth.

Hatzfeldt, Hermann Graf et al. (eds.). 1979. *Der Gorleben Report.* Frankfurt am Main, Germany: Fischer.

Heede, H.R. 1985. Invited testimony before the Subcommittee on Energy Conservation and Power, Committee on Energy and Commerce, U.S. House of Representatives (June 20). Available from the author.

JCAE (Joint Committee on Atomic Energy) Hearings. 1972a. Congress of the United States. Ninety Second Congress. Second Session. *Civilian Nuclear Power Program,* February 22 and 23. Washington, DC: U.S. Government Printing Office.

______________. 1972b. Congress of the United States. Ninety Second Congress. Second Session. *Environmental Statements 1 Through 8,* January 26, February 3, 17, 22, 23, and 29, March 1, 8, and 9. Washington, DC: U.S. Government Printing Office.

Komanoff, Charles. 1981. *Power Plant Cost Escalation: Nuclear and Coal Capital Costs, Regulations, and Economics.* New York, NY: Komanoff Energy Associates.

______________. 1980. "Power Propaganda: A Critique of the Atomic Industrial Forum's Nuclear and Coal Power Cost Data for 1978." Washington, DC: Environmental Action Foundation.

Lovins, Amory B. 1985a. "Saving Gigabucks With Negawatts." *Public Utilities Fortnightly,* 115/6 (March): 19-26.

______________. 1985b. *Least-Cost Electrical Services as an Alternative to the Braidwood Project.* Exhibit BPI-12, Illinois Commerce Commission, Docket #82-0855 (July 3). Available from the author.

______________. 1984. Testimony before the New Hampshire Public Utility Commission, Docket #84-200, 15 October.

______________. 1980a. *World Energy Strategies: Facts, Issues, and Options.* New York, NY: Harper Row Publishers, Inc.

______________. 1980b. "Nuclear Weapons and Power-Reactor Plutonium." *Nature* 283 (5750) (February 28): 817-823; Erratum, *Nature* 284 (5752) March 13, 1980: 190.

______________. 1974. "Fast Reactors: The U.S. Debate Goes Critical." *New Scientist* 61 (889) (March 14): 693-695.

Lovins, Amory B. and L. Hunter Lovins. 1984. "Prepared Testimony on Long-Term Demand for Electricity" given before the Subcommittee on Energy Conservation and Power, Committee on Energy and Commerce, U.S. House of Representatives. Ninety Eighth Congress. Second Session. February 7: 411-454. Washington, DC: U.S. Government Printing Office.

______________. 1983. Prepared testimony for the U.S. Department of Energy Public Hearing on NEPP-IV, March 2, Washington, D.C.

______________. 1982. *Brittle Power: Energy Strategy for National Security.* Andover, MA: Brick House Publishing Company.

______________. 1981. *Energy/War: Breaking the Nuclear Link.* New York, NY: Harper and Row Publishers, Inc.

Mooz, W. E. 1979. *A Second Cost Analysis of Light Water Reactor Power Plants* R-2504-RC (August). Santa Monica, CA: The RAND Corporation.

______________. 1978. *Cost Analysis of Light Water Reactor Power Plants* R-2304-DOE (June). Santa Monica, CA: The RAND Corporation.

Patterson, W. C. 1985. *The Plutonium Business.* San Francisco, CA: The Sierra Club.

Ramsay, William and Milton Russell. 1978. "Time Adjusted Health Impacts from Electricity Generation." *Public Policy* 26 (3) (Summer): 387-403.

Taylor, Theodore B. 1984. Personal communication. Formerly with the Atomic Energy Commission and currently a solar and community activist living in Damascus, Maryland.

Heede (1985) and Lovins (1985a and b) can be obtained from Rocky Mountain Institute, Old Snowmass, CO 81654.

U.S. Nuclear Exotica:
Peaceful Use of Nuclear Explosives

Richard T. Sylves

Project Plowshare, the U.S. Atomic Energy Commission (AEC) program to investigate possible non-military uses for nuclear explosives, was an offshoot of President Eisenhower's "Atoms for Peace" proposal. In a speech delivered December 8, 1953, Eisenhower declared, "this greatest of destructive forces can be developed into a great boon for the benefit of all mankind" (Hilgartner, et al., 1983: 41). After World War II and through the 1950s, many scientists, engineers, and government officials shared a vision of a nuclear utopia. Nuclear power was expected to propel aircraft, trains, naval vessels, commercial ships, rockets, and even military vehicles (Hewlett and Duncan, 1969; Hilgartner, et al., 1983; AEC, 1963: 10). Many also believed that atomic energy would be used to genetically alter crops and preserve grains, meat and fish. Moreover, nuclear reactors were envisioned which would generate huge quantities of cheap electricity (Hilgartner, et al., 1983: 43).

This scientific fascination with potential applications of atomic energy spawned a wide variety of military and civilian AEC research and development contracts, some of which were successful and many of which were not. For many advocates of nuclear technology, mechanical applications were not enough. There were those who believed that the power yielded by the thermonuclear bomb itself could have peaceful applications. Nuclear swords were to be transformed into plowshares that would dig new harbors and canals, cut gorges through mountain ridges, and open valuable mineral deposits (Hilgartner, et al., 1983: 43).

The idea for Plowshare originated at the Livermore Laboratory of the University of California in 1956. A classified symposium on peaceful use of nuclear explosives was held in February 1957 at the behest of the Livermore Laboratory. According to Irvin C. Bupp, the idea for Plowshare developed at Livermore in response to concerns resulting from the 1956 Suez Crisis. "The original notion was to use nuclear explosives to excavate a sea-level canal through Israel" (1971: 180). By June 1957, the AEC approved the Livermore plan and established Plowshare as a full-fledged program. By 1961, the AEC had formed a Division of Peaceful Nuclear Explosives, thus giving the program an institutional identity within the Commission's office hierarchy. Over the course of its programmatic lifetime, scientists at the Livermore Laboratory assumed a key role in the formulation and implementation of Plowshare tests for the AEC.

According to Brooks and Krutilla (1969), the peaceful application of nuclear explosives was not really possible until the thermonuclear bomb was developed and successfully detonated. The earliest atomic bombs, based upon uranium fission, were very expensive and released a great deal of radiation. When the first hydrogen bomb was developed, cost and radiation problems seemed to assume manageable proportions. The cost of materials for a fusion explosive, such as that embodied in the H-bomb, was far less than that for a fission explosive of equal yield. Furthermore, while radiation problems were by no means eliminated in thermonuclear blasts, fusion explosions could be tailored to very significantly reduce radiation effects (Brooks and Krutilla, 1968: 5).

Why did such a high-risk, controversial application of atomic energy intrigue the scientific community of the 1950s and 1960s? Why was Plowshare such a long-lived program, receiving federal funding from 1957 to 1978? What did the advocates of Plowshare expect it to achieve? Who and what interests were responsible for promoting Plowshare? What were their motives? What reasons can be given for the decline and eventual termination of Project Plowshare?

In answering these questions four separate, but not unrelated, test projects concocted for Plowshare are addressed. Project Chariot was to excavate a new Alaskan harbor; Project Carryall was a transportation project designed to carve a ravine out of a southern California mountain range; Project Gnome was an underground thermonuclear blast intended to serve a variety of purposes; and perhaps the largest and most controversial enterprise was the Atlantic-Pacific Interoceanic Canal Project which studied use of nuclear devices to dig a second canal. Of these four, only Project Gnome was carried through to the point of an actual thermonuclear explosion. These four projects encapsulate much of the history of Project Plowshare and serve well in answering the questions posed above.

By examining these Plowshare atomic tests, it becomes possible to identify some of the major purposes of the program. This will demonstrate that there were at least four objectives embodied in Plowshare's nuclear explosives programs, each with an associated set of interest groups. It will also reveal that there were two distinct types of Plowshare detonations. These served explicit and tangible purposes of Plowshare, as in natural resource exploitation, in transportation, and in storage or disposal of nuclear wastes.

The history of Plowshare can be characterized as a process of "policy termination." Peter deLeon indicates that in policy termination, programs undergo a process of delegitimation. Termination does not result from a single administrative or political action, but is instead the result of a long series of events and actions which culminate in the quiet cancellation of an endeavor (1978: 280-99).

Political Interests and Project Objectives

First and perhaps foremost, Plowshare was thoroughly intertwined with the issue of national security and defense. The AEC's first underground nuclear test, code named Ranier, was performed September 19, 1957. Conducted at AEC's Nevada test site located at Frenchman's Flat, Ranier was a 1.7 kiloton blast detonated in a cave that had been drilled 2,000 feet into the side of a mountain. The test was of immense military significance because it proved that nuclear tests could be carried out without producing radioactive fallout (Divine, 1978: 157-158). At the same time, Ranier opened the way to underground nuclear detonations for civilian purposes under Project Plowshare. The AEC's biannual report to Congress in 1957 openly acknowledged that Ranier, a military test, had civilian value (AEC, July 1957: 276).

Glenn Seaborg, who served as AEC Chairman from 1961 to 1971, has alleged that many military and national security interests perceived Plowshare as a way to escape the restrictions which might be imposed by a comprehensive test ban treaty prohibiting all forms of "military" nuclear bomb testing (Seaborg, 1981: 247-248). In the late 1950s and early 1960s, advocates of a test ban recognized that Plowshare could provide a means for evading restrictions against military testing. The Soviet Union had its own peaceful nuclear explosives program, which many then believed could be used in the same evasive manner. While neither the United States nor the Soviet Union would officially admit that they might exploit their peaceful nuclear programs in this fashion, each side was suspicious of the other. Seaborg indicates that Soviet Premier Khrushchev opposed any test ban treaty which permitted peaceful nuclear explosions. Khrushchev apparently believed that Plowshare-type projects could be resurrected at some future time when a system of joint international control and supervision had been established (Seaborg, 1981: 244).

Military and national security interests thus embraced Plowshare when the possibility of a comprehensive ban on weapons testing appeared a possibility. As it turned out, the Test Ban Treaty of 1963 prohibited atmospheric, surface, and underwater detonations of nuclear devices, but it did not outlaw underground testing of nuclear weapons. It is noteworthy that several military nuclear shots had Plowshare "add-on" experiments. Conversely, the seismic analysis of many Plowshare shots assisted the Advanced Research Projects Agency of the Department of Defense in testing technologies that could be used in the verification of atomic testing by other nations (AEC, 1963: 222).

Scientific advancement was a second major objective embodied in Plowshare. Nuclear explosions were, and still remain, objects of scientific curiousity. The 1950s and, to a lesser extent, the 1960s were an era of scientific exuberance. In 1945, the general public and its political representatives witnessed the effects of the first atomic bomb and were in awe of the accomplishments of the scientific and engineering community. Basic scientists - like Oppenheimer, Teller, and Fermi - who had helped create the atomic bomb were in great demand for their advice and counsel after World War II. The scientific community itself, to the extent it can be addressed collectively, saw much to be learned through the exploration and application of atomic energy in a variety of forms. Admittedly, not all scientists were advocates of Plowshare, but many believed that there was much to be discovered through Plowshare testing. Many nuclear experiments could only be performed outside the lab through an actual detonation.

In the defense-related detonation of a thermonuclear bomb in 1952, code named Mike, two new transuranic elements were identified: einsteinium and fermium. This led scientists to conclude that the large neutron flux produced in thermonuclear explosions might create "rare transplutonium isotopes and possibly, new elements by instantaneous multiple neutron capture," if the device were encased in specific kinds of materials (AEC, 1964: 171).

The Atomic Energy Commission reported in 1958 that government-supported scientists were contemplating the use of nuclear explosives as a way to produce large quantities of radioisotopes. "This might be accomplished by placing various materials around the device to be detonated in containing media so as to effect changes in the materials through nuclear reaction" (AEC, June 1958: 15). Each of the four Plowshare projects examined in this study was of high potential scientific value. Consequently, the scientific research community, including physicists, chemists, geologists, and biologists, was a source of advocacy as well as caution in Plowshare's history.

Many of these scientists worked for major AEC contract laboratories, such as Livermore or Los Alamos. There were also scientists from the Bureau of Mines, the Geological Survey, the Weather Bureau, and other agencies who were intrigued by Plowshare's potential scientific value. At least four Plowshare scientific symposia were held in the United States between 1957 and 1969. These were attended by as many as 500 to 700 people on each occasion, and were usually co-sponsored by the AEC, by major AEC labs, and by prestigious scientific and engineering societies.

A third major objective underlying Plowshare was the public works potential of nuclear explosions. Nuclear blasts intended to reshape natural surface formations were given a formal name: "geographic engineering" (Brooks and Krutilla, 1969: 36). Nuclear devices used for explosive power could redirect rivers, gouge out (or excavate) water impoundments, carve out canals, cut canyons through mountains for road and railroad building, or dredge harbors where none had existed. AEC Chairman Glenn Seaborg expected that Plowshare nuclear blasts would dig a harbor at Point Barrow, Alaska, excavate a channel across the Aleutian Island chain, deepen the Berring Straits, and accomplish other engineering feats previously beyond human capacity (Seaborg, 1981: 196). One particular remark made by Plowshare advocate Edward Teller highlights the visionary expectations of some American scientists. Teller advocated use of nuclear explosives in building a second Atlantic-Pacific interocean canal. He also envisioned using nuclear explosives to close the Straits of Gibraltar, "which would supposedly cause the Mediterranean to rise and freshen to the point that it could be used to irrigate the Sahara" (Hilgartner, 1983: 50). (The Gibraltar scheme would, however, inundate Mediterranean sea-level cities, particularly Venice.)

One organization that had a keen interest in Plowshare's public works potential was the U.S. Army Corps of Engineers. The AEC maintained that one of the ultimate objectives of Plowshare was to make possible safe nuclear explosive construction techniques which the Corps of Engineers could use in civil works projects (AEC, 1963: 228).

Legislators were enticed by the monumental constituency benefits that might stem from a project serving home states or districts. Seaborg indicates that many members of the U.S. Congress Joint Committee on Atomic Energy were ardent supporters of Plowshare (Seaborg, 1981: 197).

Commercial gain was a fourth major objective of Plowshare. Nuclear explosives could revolutionize major, grand-scale private construction enterprises. Nuclear explosives could be employed in natural resource exploitation, perhaps facilitating oil and natural gas exploration and drilling. Underground nuclear blasts could advance low-grade ore mining by stripping overburden materials. In a 1968 speech, Congressman Craig Hosmer of the Joint Committee on Atomic Energy called Plowshare "the pot of gold." He insisted that this newly invented pump would make it possible to extract $14 billion of gold, $50 billion of natural gas, and $1,000 billion of oil, oil shale, and copper, silver, molybdenum, and other metals. Hosmer even warned gold speculators that Plowshare might produce a gold glut that would cause the collapse of gold prices (Hilgartner, et al., 1983: 50-51).

The Kennecott Copper Corporation, the Continental Oil Company (Conoco), El Paso Natural Gas Company, CER Geonuclear Corporation, and the Austral Oil Company, as well as other firms, joined into formal cooperative agreements with the AEC for mining and oil and natural gas extraction experiments. The AEC even established charges that industry would have to pay in order to have the AEC detonate Plowshare explosives in commercial ventures. A two megaton Plowshare thermonuclear explosive cost $600,000 in 1964. This price included AEC arming and firing services as well as the explosive itself (AEC, 1964: 176).

Besides revolutionizing mining for hydrocarbons and valuable ores and minerals, nuclear explosives were assumed to be useful in many other commercial ventures. For example, many public and private officials believed that underground cavities formed by nuclear detonations might serve as safe repositories for the vast quantity of high-level nuclear waste produced by civilian, as well as defense, nuclear reactors. Such a use of nuclear explosives, if successful, could benefit commercial nuclear power (AEC Memo, July 6, 1972).

Types of Plowshare Detonations

The original Plowshare enterprise began with studies of the use of nuclear explosives in excavation work, but it was not long before Plowshare researchers recognized the potential value of contained underground detonations as well. In either case, the explosive is emplaced at the bottom of a large diameter drill hold. If an excavation is desired, the emplacement hole is relatively shallow and the blast fractures overlying material and excavates that material through subsidence. By varying the depth of implacement, wider or deeper craters can be obtained. For a contained blast, the emplacement hold is deeper and the explosion creates a chimney of crushed and jumbled rock, but the fractures do not reach the surface. Within the chimney porosity is very high, equal to the volume of the cavity created by the blast. If a detonation occurs at an intermediate depth so that a porous chimney of broken rock just reaches the surface, a "sink" is formed (Brooks and Krutilla, 1969: 6-7). However, the sink blasts discussed in this study are associated with excavation, rather than contained, blasts because surface effects are produced and because sink blasts were closely associated with water storage and surface impoundments, closely related to excavation projects.

The distinction between contained blasts and excavation blasts also provides an important dividing line between Plowshare detonations actually carried out, and those that were studied, perhaps planned, but never actually conducted. Beginning with Project Gnome, the first non-military detonation of a nuclear device, the AEC pursued a long series of contained underground thermonuclear explosions. While there were a number of nuclear excavation test shots conducted in Nevada and Montana, there is no public and official evidence that a Plowshare nuclear excavation shot was ever used in an actual civil or commercial construction project. The Chariot, Carryall, and the Interocean Canal excavation projects were seriously investigated and researched by the AEC in conjunction with other parties, but were never actually pursued to the point at which a thermonuclear bomb was used in a construction enterprise.

Planned Purposes of Peaceful Nuclear Explosions

Peaceful nuclear explosions were envisioned to serve three major purposes: natural resource exploitation, transportation, and storage or waste disposal.

Natural Resource Exploitation. Natural resource exploitation seemed to hold a wide array of possibilities. For example, by fracturing acquifers through nuclear detonations, large quantities of potable water might be obtained. Special measures would have to be imposed, however, to prevent contamination of this water by radiation or by substances produced by the blast. Contained blasts were expected to be the method used to fracture an acquifer.

Natural resource exploitation also encompassed the mining of low-grade ores. This was considered a possible alternative to conventional deep mine and strip mine practices. A contained blast could reveal ore deposits through block excavation or leaching. An excavation blast could be used to strip the overburden from low-grade ores. In 1967, the AEC completed a feasibility study that opened the way for the Kennecott Copper Corporation to submit a proposal to the AEC and the Department of the Interior. Designated Project Sloop, the proposal was fashioned into a plan to set off a 26 kiloton thermonuclear charge 1,200 feet underground. The site was to be Kennecott property located near Safford, Arizona (AEC, 1967: 204). The shot was expected to make 10 million pounds of copper recoverable through leaching techniques. The AEC called the radioactive contamination of copper in the project "a manageable problem" (AEC, 1966: 246).

Most significant, natural resource exploitation included the mining of hydrocarbons. Nuclear blasts were seen as a promising technique for opening impermeable natural gas or oil reserves through retorting. Moreover, they could produce the fractures needed for *in situ* recovery of oil shale. In addition, nuclear charges could heat rocks to temperatures which might make possible the recovery of fluids from tar sands.

Early in 1959, a symposium was held at the University of California Radiation Laboratory in Berkeley to consider the possibility that Plowshare's thermonuclear explosions could be used to recover oil from shale. In an article published in *Bulletin of the Atomic Scientists,* Nobel laureate Willard F. Libby discussed the prospects for mining oil and gas: A principle question was what would happen if we were to detonate, say, 300 kilotons of nuclear explosive in the Green Riven oil shale deposit in Colorado. We would expect the result to be crushed and broken shale containing 20 to 40 million barrels of oil. Gas and oil would be produced from the organic matter by the thermal energy of the explosion. The actual quantities of the products would depend upon the temperature history of the shales surrounding the explosion, the grade of the shale, and other factors. The Piceance Creek Basin of northwestern Colorado was suggested as being particularly appropriate for a test explosion of .10 kilotons. If one-half the energy released by the device were utilized in heating shale to 900 degrees (F), this would be enough to produce about 10 million cubic feet of gas and 15,000 barrels of oil (Libby, June 1959: 242).

Libby cautioned that substantial quantities of radioactive fission products could become trapped in the rock, thereby impeding efforts to extract natural gas and oil which could be safely consumed. However, conference participants, and Libby himself, advocated continued AEC Plowshare test detonations of the type then envisioned. Libby added that atomic weapons test cessation proposals, being discussed in Geneva at that time, should exclude non-military applications of nuclear explosions such as those of Plowshare. In *Kennedy, Khrushchev, and the Test Ban Treaty,* Glenn Seaborg chronicled the effort to exempt Plowshare detonations from the Test Ban Treaty negotiated between the United States and the Soviet Union. Seaborg, who served as Chairman of the AEC before, during, and long after the approval of the 1963 Test Ban Treaty, specifically referred to his hope that Plowshare projects could succeed in extracting oil from oil shales (1981: 196-198).

The Test Ban Treaty of 1963 seriously impeded Plowshare excavation blasts because of restrictions against the international migration of radioactive materials dispersed through the atmosphere by atomic detonations. Because contained nuclear explosions had little or no surface or atmospheric effects, the treaty was not a major obstacle to contained blast testing and use of nuclear explosives. For example, in 1967 plans were made to carry out the shot which Libby said was proposed as theory in 1959. CER Geonuclear Corporation, on behalf of 18 oil companies, proposed to conduct an experiment with AEC and Interior. Code named Project Bronco, the plan called for explosion of a 50 kiloton thermonuclear device in the Piceance Creek Basin 25 miles west of Rifle, Colorado. The device was to be implaced at a 3,350 foot depth and was expected to open up 1.3 million tons of oil shale to extraction operations which would yield 750,000 barrels of oil (AEC, 1967: 205). Although the Bronco shot never was carried to detonation, several related mining and gas extraction Plowshare nuclear devices were fired in the years after 1963.

Records demonstrate that as late as 1973 the AEC was studying two more mining-related nuclear experiments. From 1947 to 1975, the AEC regularly solicited the advice of its most prestigious advisory committee, the General Advisory Committee (GAC). On February 15, 1973, the committee took up the matter of Plowshare. GAC Chairman Howard G. Vesper, an executive of Standard Oil of California, asked the committee to comment on the reduction in spending for Plowshare contained in the AEC's fiscal year 1974 budget. Vesper lamented the cuts and praised Plowshare. He mentioned a 1972 GAC review of the project which concluded, "it has potential for increasing critical energy supplies in a reasonable time schedule" (GAC Minutes, 1973: 18). Lombard Squires of E. I. du Pont de Nemours & Company, a firm with a long history in the field of atomic energy, suggested that the AEC proceed with two planned Plowshare shots: Rio Blanco and Wagon Wheel. But he added that the program might be impractical thereafter. Stanford University's Hubert Heffner suggested that Plowshare was at a point where industry should pursue it rather than government alone.

Committee member James H. Sterner, of University of Texas, said that the AEC should be prepared to undertake nuclear shots but he doubted that "any geographical area" would submit to such events. To this, Livermore Laboratory's Michael M. May responded, "local personages support AEC programs whenever they are informed in a straightforward manner" (GAC Minutes, 1973: 19). May stated that opposition was usually from organized groups residing elsewhere. May also argued that industry could not develop the nuclear devices, that funds cut from Plowshare should be restored, and that Plowshare was "an excellent bargaining chip in negotiations with oil- and gas-rich nations." Vesper agreed with May. On the matter of local test impact of underground thermonuclear bomb detonations, Vesper said, "the public need expect some inconvenience if it is to secure its energy requirements" (GAC Minutes, 1973: 19).

Vesper then asked if anyone on the committee objected to recommending that Plowshare device development be adequately funded. Squires answered affirmatively but suggested that the AEC explore possible private industry funding of the project. May observed that industry was not certain it could legally engage in the project; moreover, security restrictions would prevent industry from knowing what exactly it was buying in its participation. The GAC ended discussion of Plowshare at this 1973 meeting, noting that in its 1972 report the committee had endorsed Plowshare as a possible way to create repositories for long-term storage of nuclear wastes (GAC Minutes, 1973: 19).

Storage and Waste Disposal Structures. Many Plowshare proponents believed that nuclear explosives could be used to create underground water reservoirs or storage caverns, including the construction of dams, spillways, or drainage craters. Excavation or sink blasts would have to be the methods for creating the necessary surface structures or formations. Contained, excavation or sink blasts might be employed to fashion reservoirs or storage caverns. Water storage problems include contamination, leakage, and slope stability (Brooks and Krutilla, 1969: 30). For a number of years the Bureau of Mines, the Geological Survey, and other federal agencies investigated nuclear explosions as an instrument in water supply programs. In 1968 the Governor of Arizona approached the Interior Department with a plan to engage in cooperative studies "of the feasibility of applying nuclear explosions to water resource management in his State" (AEC, 1968: 202). This proposal was dubbed Project Aquarius. There is no evidence, however, that Aquarius was pursued beyond the research stage.

An additional storage purpose that received scientific and industry attention for many years was the creation of storage caverns for natural gas. The AEC and the Columbia Gas System Service Corporation pursued Project Ketch in the mid-1960s. This proposal originally called for the discharge of a 24 kiloton nuclear explosive at a depth of 3,000 feet in impermeable shale at a site 12 miles southwest of Renovo, Pennsylvania. The cavity created was expected to store under high pressure 465 million cubic feet of natural gas (AEC, 1968: 202). This endeavor was hampered by economic, not technical, obstacles. According to Brooks and Krutilla, gas line companies have found it economical to store gas obtained in excess of consumption during the summer months, so that they could meet the increased demand of winter months. Because producing areas are usually far from major consuming areas, it is important for the natural gas industry to store gas as close to the areas of consumption as possible. This meant that nuclear caverns as gas storage repositories should ideally be sited close to the large eastern gas-consuming market (Brooks and Krutilla, 1969: 35). This reasoning was apparent in Project Ketch's Pennsylvania location. This is where the appeal of nuclear caverns seems to have waned. The problems of detonating underground nuclear devices in a highly populated region with numerous buildings and man-made structures were more serious than the problems encountered in use of non-nuclear alternatives. For example, natural underground caverns might be used, despite their relatively great distance from eastern markets. Furthermore, given the limited dimensions needs for storage caverns, conventional explosives seemed superior to nuclear.

Among all storage purposes, nuclear caverns created for disposal purposes seemed to get the most sustained attention. These underground cavities could be employed for long-term storage of toxic wastes or radioactive effluents if impermeability could be insured (Brooks and Krutilla, 1969: 31).

Transportation. The third and most controversial category of Plowshare's major purposes was that of transportation. According to the AEC, Plowshare had civil engineering applications because "large volumes of earth and rock can be moved or broken" by "nuclear explosives at costs below those of conventional methods" (AEC, 1958: 51). The Commission's 1958 Report to Congress talked of nuclear explosives for use in digging harbors and canals, clearing navigation hazards, and for a variety of other transportation-related tasks.

Besides the transportation projects examined below, thought was given to several others. In 1964 the AEC cooperated with the Tennessee-Tombigbee Waterway Development Authority, the Army Corps of Engineers, and other agencies in a Plowshare-related scheme. These agencies evaluated "the technical feasibility of using nuclear explosives to excavate approximately 3 miles of the 'divide cut' through low hills in the Northeast corner of Mississippi (the highest terrain) on the proposed waterway" (AEC, 1964: 165). Ultimately, nuclear explosives were determined to be inappropriate for this canal project. In 1969 the Australian government engaged the AEC in the study of a project to develop a harbor at Cape Keraudren through nuclear excavation. This project was abruptly canceled when the mining company which initiated the scheme withdrew interest in the venture (AEC, 1969: 202).

The key problem of nuclear explosives in transportation is that every detonation would have to be of the excavation type. This meant that the thermonuclear device would have surface and atmosphere effects which could pose dangers in the form of groundshock, air blast, and radioactive fallout. In the case of air blast, waves set up from near surface explosions not only would move outward from the explosion but might also be reflected from, and amplified by, layers in the atmosphere. This could produce effects several hundred miles away from the shot epicenter. Groundshock would mean earthquake-like tremors that result from the underground detonation of large explosive charges. The magnitude of groundshock is a function of both the depth of implant and the natural geology of the site. In excavation shots fallout may be distributed over an area of several hundred square miles or more. Finally, excavation-type detonations would involve release of highly radioactive isotopes, including strontium-90 and cesium-137 (Brooks and Krutilla, 1969: 12). As a result, excavation shots might require the evacuation of population and the reinforcement of adjacent buildings and structures.

From the inception of Plowshare until its termination, constant efforts were made to reduce the amount of radioactivity generated in each nuclear explosion. For example, the Palanquin nuclear test explosion in 1965 was designed to study how implacement techniques might reduce radioactivity release. The AEC expressed disappointment with this test, as it would with many others, because results of this excavation blast revealed that more radioactivity was released than expected (AEC, 1965: 203).

Four Plowshare Case Studies

The four projects discussed here illustrate the range of purposes envisioned for Plowshare, the interest groups that supported it, the kinds of blasts planned (if not executed), and the problems encountered. Project Gnome was an underground blast intended to answer a number of questions, some scientific and some commercial. Project Chariot was to excavate a new Alaskan harbor, thus opening a new region for mineral resource development. Project Carryall was a combined road/railway transportation project that was to carve a ravine out of a southern California mountain range. The Atlantic-Pacific Interocean Canal Project - the largest and most controversial enterprise - studied the possible use of nuclear devices to dig a second isthmian canal. It was expected to have political, military, and commercial implications. Of these four projects, only Gnome was carried to the point of an actual thermonuclear explosion.

It is important to acknowledge the number and years of Plowshare experiments which actually culminated in thermonuclear explosions. There were five Plowshare blasts before the October 10, 1963, approval of the Partial Test Ban Treaty. One was conducted in 1961 (the Gnome experiment), three in 1962, and one more in early 1963. In the period 1963 to 1968, after the treaty restrictions went into effect, there were 23 more Plowshare detonations up through 1968. One shot was fired in late 1963, six in 1964, one in 1965, four in 1966, three in 1967, and eight in 1968. The figure of eight in 1968 includes five devices separately used in the Project Buggy test (Williams and Cantelon, 1984: 242). Plowshare detonations after 1968 were the Rulison shot of 1969 and the Rio Blanco shot of 1973.

Project Gnome

Project Gnome was one of the first Plowshare program experiments. Preliminary Gnome feasibility studies began before 1958. In 1960, the plan called for detonation of a 10 kiloton explosive at a depth of 1,200 feet (AEC, 1960: 153). By 1961 this had been scaled back to a 5 kiloton explosive. Gnome would be used to study the production and recovery of heat and isotopes produced in a contained nuclear explosion (AEC, 1961: 210).

AEC officials believed that the energy stored as heat in the nuclear cavity might be released in a controlled manner by use of transfer agents such as water, carbon dioxide, or nitrogen. The energy recovered from this heat could be used "industrially or in the production of electric power" (AEC, July 1958: 52). In other words, the tremendous heat energy stored in the nuclear cavern might be exploited on the surface through various kinds of power production facilities.

Project Gnome enabled AEC-supported laboratories - such as Lawrence Livermore, Los Alamos Scientific Lab, and Brookhaven National Lab - to conduct neutron experiments that were "either impossible or extremely difficult to perform" with then-conventional laboratory techniques (AEC, 1960: 153). In a related sense, the AEC looked upon Project Gnome as a possible means for producing man-made isotopes. The AEC hoped to show that Gnome was an economic alternative to standard methods of power and isotope production. Economical isotope production had important implications for national defense and security because it could facilitate production and fabrication of nuclear warheads of various types.

When Gnome was exploded in December 1961 in New Mexico, the ground surface heaved upward producing a permanent two-foot bulge. The explosion, however, did not reach the surface. Radioactive steam produced from heating water in the salt bed cavity was vented to the air above the site through a shaft. Maximum 10,000 roentgen per hour readings in downwind monitoring caused the AEC to close several public roads near the site (AEC, 1961: 212-213).

When President Kennedy approved the Gnome shot in October 1961, he said it was "a further example of this country's desire to turn the power of the atom to man's welfare rather than his destruction" (AEC, 1961: 210-211). Project Gnome was to open the way to a succession of contained nuclear explosions in the 1960s. The large underground cavity created by the Gnome blast was opened to press tours by the AEC. Among a number of Project Gnome findings was that production of electric power from nuclear excavated salt cavities was economically uncompetitive with alternative electric power production technologies (AEC, 1962: 254).

Project Chariot

In 1958 the Atomic Energy Commission announced that it was pursuing a detailed investigation to determine whether nuclear explosives could be used to excavate a harbor in northwestern Alaska between Cape Seppings and Cape Thompson. This area is located north of the Arctic Circle. The AEC insisted that a harbor in this region was necessary in order to make possible the development of mineral resources (AEC, June 1958: 15). Its survey of the area estimated costs, collected physical and economic data, and evaluated safety factors. The AEC also notified Canada and the U.S.S.R. about the Alaska harbor studies.

Project Chariot bioenvironmental and engineering field surveys were conducted in 1959. The Fish and Wildlife Service, the Weather Bureau, the Public Health Service, the Army of Engineers, and scientists from several universities performed much of this work (AEC, December 1959: 93). Two years later Project Chariot research continued, but the AEC had not yet received an authorization to explode a nuclear device. Nevertheless, the AEC had formulated the project's civil engineering plan.

This plan called for simultaneous detonation of four 20 kiloton explosives and one 200 kiloton nuclear explosive. The smaller charges were to be buried at a depth of 400 feet and the large charge at an 800 foot depth. These explosives were to create a channel 900 feet wide, with a 200 foot maximum depth and a 2,000 foot length. The large nuclear charge would excavate an 1,800 foot diameter, 400 foot deep turning basin, which would link the channel created by the smaller nuclear charges to the ocean (AEC, 1961: 215-216).

By 1962 AEC's environmental studies of Project Chariot's predicted effects concluded that harm to Eskimos and to the Alaskan environment was "exceedingly remote" (AEC, 1962: 253). Despite years of research, however, the AEC was never given permission to detonate a nuclear explosive at the Chariot site. The project was bypassed as the AEC pursued numerous nuclear excavation shots at the AEC Nevada Test Site in order to perfect use of nuclear explosives. For example, in 1963 Project Buggy yielded information about what would happen if five nuclear explosives were set off at the same time to produce a ditch. This was followed by an estimated 100 kiloton shot in hard rock, another nuclear shot in sandstone, and by a multiple charge shot in hard rock with varying surface terrain (AEC, 1962: 250-251).

A major reason why Chariot was abandoned stemmed from the prohibitions imposed by the 1963 Test Ban Treaty, which banned atmospheric detonations of nuclear explosives. There were specific provisions that applied to tests producing debris or fallout which could not be contained within the borders of the nation exploding the mechanism. There was little assurance that fallout produced by the

Chariot explosions could be completely confined to the United States. By 1963 Project Chariot was relegated to a condition of indefinite postponement, and the AEC began returning test site land to the public domain (AEC, 1963: 217).

While it may seem that Project Chariot was of little consequence, appearances deceive. Project Chariot laid the foundation for numerous domestic U.S. nuclear excavation test shots conducted in a number of western states. Because of the attendant risks attached to the Chariot shots, AEC scientists did not favor firing the Chariot devices for research purposes; instead, they wanted the Chariot shots to be based on predictable, routine, and safe pre-testing. Furthermore project Chariot co-existed with, and to a degree set the stage for the most ambitious Plowshare project, nuclear excavation of a second isthmian canal.

Atlantic-Pacific Interocean Canal Project

Why a second Panama Canal? A second and deeper canal could accommodate more vessels and vessels of larger size. It would presumably be less vulnerable to military attack, because as a sea-level canal it would have no lock systems. It was alleged that such a canal would be simpler to run and would have low operating costs. Proponents argued that it might even reduce political tensions between Panama and the United States (Brooks and Krutilla, 1969: 39).

At the Second International Conference on Peaceful Uses of Nuclear Energy held in Geneva in 1958, Edward Teller said that Plowshare could help to dig a sea-level canal across Panama (Hilgartner, 1983: 50). Two years later, George B. Kistiakowsky, President Eisenhower's scientific adviser at the time, made the following entry into his diary: From a remark of the President, I gather he intends to address the UN General Assembly and at that time (September) announce that we are going ahead with the Plowshare program for digging a second Atlantic-Pacific canal through Mexico, which has agreed to the project. It was obvious that both the President and the Vice-President think Plowshare is terribly important (1976: 365). Kistiakowsky's entry has a subscript which disclosed that Eisenhower had been "premature" in his reference to a canal through Mexico "which had not formally agreed to this project" (1976: 365).

In 1962, legislation was introduced in Congress to authorize the Panama Canal Company to conduct surveys of possible routes for the construction of a sea-level, trans-isthmian canal. This was welcomed by the AEC, which was continuing to pursue nuclear excavation test shots which could serve the canal project (AEC, 1962: 263). In September 1964, Public Law 88-609 was enacted. This measure authorized the president to appoint a five-man commission which was to investigate and study the feasibility of, and the most suitable site for, the building of a sea-level canal connecting the Atlantic and Pacific oceans. This commission was to determine the best means of constructing the canal, "whether by conventional or nuclear excavation, and the estimated cost of each" (AEC, 1964: 164-165). It was not until April 18, 1965, that President Johnson made the five appointments to the Canal Study Commission. By this time, the three preferred routes were the present Canal Zone, the Darien Region of Panama well east of Panama City, and northwestern Columbia astride the border of Panama. Permission to conduct on-site surveys of Route 17 (Sasardi-Morti) in the Darien region and Route 25 (Atrato-Truando) through Columbia were secured by the United States from Panama and Columbia, respectively, in 1966 (AEC, 1966: 252).

In anticipation of use of nuclear explosives in excavation of a new interocean canal, the AEC planned a series of nuclear excavation test shots at its western U.S. facilities. The largest shot by far was called Project Sedan. It was a 100 kiloton detonation set off 635 feet below a central Nevada site. The crater formed was 1,280 feet deep. This test occurred in July 1962, before the successful negotiation of the partial Test Ban Treaty. A second 100 kiloton cratering experiment was deferred and later scaled back as a result of limitations imposed by the 1963 treaty (AEC, 1963: 210-213).

Ratification of the treaty forced the AEC to move to lower-yield nuclear shots (smaller than Project Sedan) less likely to release high amounts of radioactive matter to the atmosphere (AEC, 1963: 210). With the firing of Project Sulky in 1964, the AEC concluded that, with the proper design, excavation experiments could be carried out without causing radioactive debris to move beyond national boundaries (AEC, 1964: 159).

By 1966 the AEC initiated Project Cabriolet, a nuclear excavation test specifically tailored to determine the feasibility of using nuclear explosives in building a sea-level canal across the Pan-American Isthmus. Cabriolet was a 2.5 kiloton explosion which was set off in January 1968. It was 40 times smaller than the Project Sedan shot, but nonetheless produced a crater 120 feet deep and 360 feet in diameter (AEC, 1968: 196-197). Project Buggy was one of the first nuclear row charge experiments. It too was detonated in 1968. The five nuclear charges were spaced 150 feet apart and 150 feet underground. They produced a channel 300 feet wide, 900 feet long, and 80 feet deep (AEC, 1967: 206). The AEC reported that some radioactivity migrated from the site of each of these tests, but no further details were given.

As the Atomic Energy Commission continued to gather field data from the two proposed route areas in Central America, and as the agency continued to assess the results of its U.S. tests, the AEC announced that cratering experiments demonstrated "that nuclear excavation of a sea-level canal is a possible means of constructing such a waterway" (1968: 198).

In its recommendations concerning nuclear excavation of a second Panama Canal, the Canal Study Commission 1970 report offered a contrary view. If Route 17, the Sasardi-Morti route, had been selected as the canal site, this would have required breaking through a "backbone-like ridge that reaches 1,100 feet in elevation at one point" (Brooks and Krutilla, 1968: 39). Cuts of this magnitude had never been attempted. The Study Commission calculated that Route 17 would require 250 separate nuclear excavation charges which would yield a cumulative 120 megatons of explosive power. This would have been accomplished with the firing of 30 salvos of varying yields over at least a three-year period. The largest salvo would have a yield of 11 megatons (Canal Study Commission, 1970: 37). Route 17 nuclear excavation across Panama would have effected 6,500 square miles and an estimated 43,000 inhabitants would have to be evacuated for a period of months. The effects of ground shock and air blast might be felt as far as 300 miles away from the site, with glass breakage and masonry damage extending from Panama City to the Columbian border area.

A similar continental divide cut was required for Route 25, or the Atrato-Truando route across northwestern Columbia. About 150 individual nuclear charges would be fired in an estimated 21 row salvos over a three-year period. The largest salvo would be 13 megatons and the total explosive power of all of the Route 25 nuclear shots was predicted to be about the same as for Route 17: 120 megatons. Route 25 was expected to affect 3,100 square miles which encompassed an approximate 10,000 inhabitants. As in the Route 17 plan, these people would need to be evacuated for many months (Canal Study Commission, 1970: 40). Route 17, by comparison, affected a far larger area and more than four times as many people.

Nuclear explosives were to excavate only a fraction of the path of each canal route. Nevertheless, the ground motions and radioactive fallout produced in this work would have had serious implications for the host country. According to Weisner and York, the cumulative nuclear explosive energy released in canal excavation "would be twenty times the cumulative amount of explosives used in all previous wars combined" (1964: 28). The Canal Study Commission reported that many scientists around the world were alarmed about the radioactivity that would be released to the environment in nuclear excavation.

The Canal Study Commission concluded that the Test Ban Treaty of 1963 would have to be modified to permit major nuclear excavation projects. This was because "there seems to be no possibility of excavating an Isthmian Canal with nuclear explosives without transport of some radioactive material across territorial boundaries" (Canal Study Commission, 1970: 43-44). Moreover, the group's 1970 report indicated that each prospective host-country opposed nuclear canal excavation even more vehemently than in 1964.

The Canal Study Commission's 1970 report also referred to serious economic dislocations caused for each route by nuclear excavation. For example, Route 17 would have an impact upon large numbers of Choco and Cuna Indians, who have strong cultural attachments to their lands. Route 25, while affecting far fewer people, also would uproot Choco Indians in northwest Columbia (Canal Study Commission, 1970: 44-45). Routes 17 and 25 both would drain financial resources and economic opportunity away from the existing Canal Zone area of Panama.

The Canal Study Commission disclosed that the AEC had only been able to conduct three U.S.-based nuclear excavation test shots, instead of the original planned minimum of six. Throughout the 1960s, U.S. Plowshare excavation testing grew increasingly controversial. Consequently, Congress declined to furnish all of the funds needed to prepare each canal-related test shot. Presidents Kennedy, Johnson and Nixon were not always forthcoming in granting prompt permission to fire each Plowshare excavation shot. AEC Chairman Seaborg recommended to the Study Commission that because "the AEC had not been able to do all of the experiments required to make a determination of the feasibility of using nuclear explosions" in canal excavation, "any decision in the near future to construct a sea-level canal would have to be made without reliance on nuclear excavation" (AEC, 1970: 201-202).

In the end, nuclear excavation of a second canal was deemed to be a premature use of the technology. The Study Commission's Technical Associates strongly advised against nuclear canal building on the grounds that appropriate techniques for such an enterprise would not be available within the next 10 years (Canal Study Commission, 1970: 40). The new canal proposal itself fell by the wayside as the United States and Panama entered into a long process of negotiation that ultimately resulted in transferral of the Panama Canal to the Republic of Panama.

Project Carryall

Project Carryall was another highly controversial nuclear excavation proposal. In 1963, executives of the Santa Fe Railway asked AEC officials about the possibility of using nuclear explosives to excavate a cut through the Bristol Mountains north of Amboy, California. This site was in the southeastern portion of the state at a point equidistant between Barstow and Needles. The railroad expected that the project would shorten travel time, reduce fuel consumption, and preclude construction of a long, expensive tunnel (Brooks and Krutilla, 1969: 36).

Soon after the Santa Fe Railway made its proposal, it was joined by the California Division of Highways. The division was then transforming Highway 66 into Interstate 40, which was under construction at the time. The relocation plan called for a northern, higher, and straighter route, which the Carryall Project cut could provide. Therefore, both the railroad and the highway authorities could use the 350 foot deep cut created by the blast (AEC, 1963: 227).

A Carryall feasibility study was completed in late 1963 by a study group composed of staff members from the AEC San Francisco Operations Office, the Lawrence Livermore Laboratory, the Santa Fe Company, and the California Division of Highways. The proposed pass would be about 10,000 feet in length, would involve cuts up to 350 feet deep, and would dig out about 60 million cubic yards of rock. The study concluded that the project was technically feasible (AEC, 1963: 227).

Carryall was not realized "because it ran afoul of the time schedule for completion of the Interstate Highway System" (Brooks and Krutilla, 1969: 36). By 1965, the AEC reported that Carryall was largely defunct (AEC, 1965: 196). The Carryall proposal, according to Brooks and Krutilla, would have been of most economic gain to the railroad (1969: 37). The project involved an unusual assortment of public and private parties who were encouraged by the AEC to explore the feasibility of use of atomic explosives.

The Decline of Plowshare

Many of the seeds of Plowshare's demise were planted by the emergence of the U.S. environmental movement. This was even intimated in AEC's 1970 Report: Along with the growing concern for environment, there have been an increasing number of indications of public uneasiness about nuclear testing and particularly about high-yield nuclear testing and Plowshare excavation experiments. Because these matters had become worrisome to the public, a considerable bioenvironmental study program has grown up in the past years and this program was intensified during 1970 (AEC, 1970: 42). What this statement omitted was that even Plowshare underground nuclear testing triggered public fears and consternation.

The political implications of this public concern were felt all the way to the White House. By 1970, all of AEC's plans for nuclear testing were subject to review by the National Security Council's undersecretaries committee. Furthermore, the Council on Environmental Quality, a statutory executive office, examined all Plowshare testing plans for compliance with the National Environmental Policy Act of 1969 (NEPA). Environmental impact statements were to be prepared for each planned Plowshare explosion (AEC, 1970: 42).

The AEC became highly vulnerable to accusations by public interest science organizations that full public disclosures about planned Plowshare detonations were not being made. As more information about each shot became publicized, more of the public became concerned about proposed shots. Civilian use of nuclear explosives in Plowshare could no longer proceed as elite and concealed policy making. NEPA could not force the declassification of Plowshare thermonuclear technical data that described how the devices themselves functioned. The on-site and off-site consequences of a Plowshare explosion, however, were subject to public disclosure through legally mandated environmental impact statements required under NEPA.

The AEC's final official mention of Plowshare as an operating program came in its 1974 Annual Report to the Congress. Reference was made to the Gasbuggy detonation of 1967 and the Rulison blast of 1969. The AEC reported that it was continuing to assess the results of its 1973 Rio Blanco underground thermonuclear explosion. Rio Blanco proved to be the final detonation of a nuclear device for

Project Plowshare. The Kennecott Corporation continued analyzing the potential feasibility of a Project Wagon Wheel shot, but the firm never approached the government with a request to co-sponsor the project through an actual detonation.

In the mid-1970s the AEC itself was undergoing a transformation which would precede its bifurcation into the Nuclear Regulatory Commission and the Energy Research and Development Administration. In this final phase, AEC moved its applied technology projects away from peaceful nuclear testing and toward "environmental research, energy storage, and transmission systems, synthetic fuels and non-nuclear energy" (AEC, 1974: 26).

It was not possible for the AEC to carry Plowshare into an era of renewable energy, energy conservation, and environmental sensitivity. In 1971, the AEC had participated in a joint government-industry feasibility study which concluded "there is no technical reason why electricity cannot be produced from geothermal energy sources with the aid of nuclear explosives" (AEC, 1971: 142-143). Yet, this and other attempts to creatively move Plowshare into an era filled with alternative energy schemes were unsuccessful.

The mixed results and public controversy produced by the 1973 Rio Blanco gas stimulation experiment in northwest Colorado helped close the door on use of nuclear explosives as instruments to be used "in augmenting the Nation's energy resources" (see AEC, 1971: 139).

Conclusions

Plowshare was, in a sense, two separate programs. One was for nuclear excavation projects applied to grand-scale civil engineering ventures. Much of what had sustained nuclear excavation Plowshare in the 1960s was the hope and belief that this new instrument of civil engineering would prove its value in construction of a second great Pan-American canal. The other was for contained underground blasting to serve parties interested in mining, underground natural gas storage, and long-term disposal of toxic and radioactive materials. Both programs were intertwined with military and national security-related experiments. Given the heavy security which justifiably surrounded AEC use of thermonuclear devices, and fears of nuclear terrorism as well as nuclear proliferation concerns, even Plowshare's most ardent supporters never expected the project to hand private industry a thermonuclear explosive device.

The Plowshare Project captured the imagination of many scientists, engineers, and elected political officials. Many expected Plowshare to be a tool of geographic or planetary engineering, but "excavation" Plowshare failed to meet political and diplomatic restrictions due to the surface and atmospheric effects of nuclear excavation explosions. Efforts to down-size Plowshare excavation blasts to comply with treaty restrictions moved nuclear explosives into a range in which more economical conventional explosives were the preferred alternative. This alone, however, was not the cause of Plowshare's demise. Environmentalists expressed criticism of Plowshare excavations' ecological effects. It became difficult enough to win approval for major public works construction projects that did not involve nuclear explosives. Those planning nuclear excavation faced a barrage of public and private criticism.

"Contained-blast" Plowshare seems to have been exhaustively researched, particularly for its mining potential. Major oil and natural gas companies, as well as a huge copper mining firm, pursued tests and feasibility studies with the AEC up to and through actual AEC-directed underground nuclear detonations. The cost of removing radioactive contaminants from oil, natural gas, copper, water, or other substances of commercial value, however, made nuclear fracturing or nuclear cavern creation a poor business investment with high risk. If business was to become a co-partner in the use of nuclear explosives, as envisioned, new and extraordinary responsibilities would have to be shouldered. The federal government provided no insurance against damage claims put forward by third parties affected by Plowshare blasts. There was no protection against liability, as in the case of the Price-Anderson Act of 1957 for commercial nuclear (reactor) power.

Where "excavation" Plowshare failed to meet the restrictions of international relations in a political realm, "contained-blast" Plowshare failed to prove practical and profitable in an economic realm. Both dimensions of Plowshare were forced to deal with the problem of political and legal accountability. The U.S. Government was held increasingly accountable for the international consequences of nuclear excavation shots. The AEC and its industrial co-partners were held increasingly accountable under property, negligence, and environmental protection laws for deleterious effects produced in contained nuclear explosions. Despite significant technical advances in the design of thermonuclear devices, Plowshare nuclear explosions could not be engineered to meet constantly tightening sets of restrictions.

Peter deLeon indicates that many programs and policies can fend off termination efforts if those committed to the program have a great intellectual reluctance to see it end; if there is a high degree of institutional endurance which keeps a policy moving forward; if the organization implementing the policy is able to make dynamic and defensive organizational adjustments; if an anti-termination coalition can be forged to block threatened termination actions; if a body of law or other legal obstacles prevent dissolution of a program; or if there are considerable sunk costs in a program which will be lost in termination (1978: 280). Plowshare was terminated because it failed on virtually all of these counts. The scientific community, one of the program's staunchest supporters at the beginning, gradually lost interest in using thermonuclear explosions to pursue scientific experiments. The AEC Division of Peaceful Nuclear Explosives lost funding and personnel over the years and so there was not a strong internal group of AEC officials fighting to save Plowshare in the 1970s. The dissolution of AEC in 1974 thrust the program into ERDA, a multiple-mission energy agency which relegated atomic energy programs to a diminished position in the agency. Consequently, ERDA was not predisposed to save Plowshare. The effort to use Plowshare in geothermal energy research represented a feeble attempt at defensive adjustment against proposed termination. There were no anti-termination coalitions fighting to save Plowshare in the 1970s because the business community became disenchanted with Plowshare and its marginal commercial benefits. The defense establishment did not need Plowshare because the military could have its underground tests conducted by the AEC, ERDA, or the Department of Energy unencumbered by major treaty restrictions. The Joint Committee on Atomic Energy, an early and long-time source of Plowshare protection and promotion was disbanded in the mid-1970s. Plowshare had no firm basis in federal statutes, other than those which promoted the general application of atomic energy for civilian uses in the Atomic Energy Act of 1954. Because Plowshare progressed as an R&D project pursued with AEC contract lab staff that could be easily detailed to other project work, and because Plowshare did not rely on or generate major physical facilities, there were few sunk costs which the government would lose if the program were canceled. The official termination of Plowshare came as a budget decision of the Congress, a decision unopposed by the President.

The story of Plowshare is the story of sustained and futile scientific and engineering optimism in the face of a world public that became increasingly intolerant of such nuclear ventures. The base of nuclear energy's constituency and private clientele support eroded significantly in the 1970s and Plowshare was too marginal to maintain. The termination of Plowshare was a reluctant admission that a nuclear utopia was not imminent.

REFERENCES

AEC (U.S. Atomic Energy Commission). 1974. *Annual Report to Congress.* Washington, DC: U.S. Government Printing Office.

_______________. 1973. *1973 Annual Report to Congress: Operating and Developmental Functions* Volume 1. Washington, DC: U.S. Government Printing Office.

_______________. 1972. *1972 Atomic Energy Programs: Operating and Developmental Functions* Volume 1. Washington, DC: U.S. Government Printing Office.

_______________. 1959-71 (inclusive). *Major Activities in the Atomic Energy Programs* Annual Volumes. Washington, DC: U.S. Government Printing Office.

_______________. 1958. *Research on Power from Fusion* Volume 24, January-June. Washington, DC: U.S. Government Printing Office.

_______________. 1957. *Progress in Peaceful Uses of Atomic Energy* Volume 23, July-December. Washington, DC: U.S. Government Printing Office.

Brooks, David B. and John V. Krutilla. 1969. *Peaceful Use of Nuclear Explosives: Some Economic Aspects.* Baltimore, MD: Resources for the Future, distributed by The Johns Hopkins University Press.

Bupp, Irwin C. 1971. "Priorities in Nuclear Technology: Program Prosperity and Decay in the United States Atomic Energy Commission, 1959-1971." Unpublished Ph.D. dissertation. Cambridge, MA: Harvard University.

Canal Study Commission (Atlantic-Pacific Interocean Canal Study Commission). 1971. *Interocean Canal Studies.* Washington, DC: U.S. Government Printing Office.

deLeon, Peter. 1978. "A Theory of Policy Termination." In Judith V. May and Aaron Wildavsky (eds.), *The Policy Cycle.* Beverly Hills, CA: Sage Publications.

Divine, Robert A. 1978. *Blowing on the Wind: The Nuclear Test Ban Debate 1954-1960.* New York, NY: Oxford University Press.

Hewlett, Richard G. and Francis Duncan. 1969. *Atomic Shield 1947/1952* Volume 2. University Park, PA: Pennsylvania State University Press.

Hilgartner, Stephen, Richard C. Bell, and Rory O'Connor. 1983. *Nukespeak: The Selling of Nuclear Technology in America.* New York, NY: Penguin Books.

Kistiakowsky, George B. 1976. *A Scientist at the White House.* Cambridge, MA: Harvard University Press.

Libby, Willard F. 1959. "Nuclear Energy - Some New Aspects." *Bulletin of the Atomic Scientists* 15/6 (June): 240-244.

Seaborg, Glenn T. with Benjamin S. Loeb. 1981. *Kennedy, Khrushchev, and the Test Ban.* Berkeley, CA: University of California Press.

Weisner, Jerome and Herbert York. 1964. "National Security and the Nuclear Test Ban." *Scientific American* 211/4 (October): 27-35.

Williams, Robert C. and Philip L. Cantelon (eds.). 1984. *The American Atom.* Philadelphia, PA: University of Pennsylvania Press.

Technology, Politics and Ideology:
The Vicissitudes of Federal Solar Energy Policy, 1974-1983

Eugene Frankel

Between 1974 and 1983 the federal government spent almost $3 billion in research, development and demonstration on solar technologies. In addition, more than a billion dollars were spent on tax credits and other financial incentives. These totals, however, conceal a remarkably stormy history. Annual appropriations varied by as much as a factor of ten over the period; organizational changes occurred almost every other year; the philosophy guiding the programs was radically revised at least twice. Underlying these changes were profound differences in approach to energy research and development policy among key decision makers in the federal government: the Congress, the scientific and technical agencies, and the executive offices of two very dissimilar administrations. This paper examines four approaches to national solar energy research and development policy that characterized the principal decision makers in this era: the congressional approach, as embodied in the National Solar Heating and Cooling Demonstration Act of 1974; the "ERDA approach," as exemplified by the first years of the national solar energy research and development program; the Carter administration approach, characterized by the 1978 Solar Domestic Policy Review; and the Reagan administration approach, as manifested in its first two proposed Department of Energy solar budgets. I will attempt to show, for each of these approaches, how the interplay of technology, politics and ideology led to outcomes that were quite different from the expectations of the policy makers and not always consistent with the overall national interest.

Solar R&D: An Overview

In December 1973 the House Committee on Science and Astronautics (re-named the Committee on Science and Technology on January 3, 1975) marked up and passed on to the full House a bill numbered H.R. 11864, the "Solar Heating and Cooling Demonstration Act of 1973," which later became Public Law 93-409 (CRS, 1978: 203). In the years that followed solar energy became one of the most important and controversial issues in national domestic policy. More than 367 bills related to solar energy were introduced in Congress between 1973 and 1980; 53 of those passed (Moore, 1983). More than 122 bills concerned exclusively with solar were introduced; 15 of those passed. The federal government spent close to $3 billion in research, development and demonstration on solar technologies between 1974 and 1983. More than a billion dollars were spent on tax credits and other financial incentives.

By 1983 at least eight solar technologies were actively competing in the energy marketplace: (1) solar water heating, (2) passive solar design for home heating, (3) active solar space heating, (4) wood stoves, (5) alcohol fuels, (6) low Btu gasification and direct combustion biomass systems, (7) low-head hydroelectric dams, and (8) small wind machines and crystalline silicon photovoltaics modules for remote and third world applications. Another seven were in the engineering demonstration or early commercialization phase: (1) parabolic trough concentrators, (2) solar thermal central receiver systems, (3) large multi-megawatt wind turbines, (4) amorphous silicon photovoltaic modules, (5) medium Btu biomass gasification systems, (6) on-farm anaerobic digestion systems, and (7) passive solar cooling and daylighting for commercial buildings. Eight more were still in the research and development phase: (1) thin film polycrystalline and silicon ribbon photovoltaic cells, (2) high efficiency concentrating photovoltaic systems, (3) biological production of hydrogen, (4) ocean thermal energy systems, (5) parabolic dish concentrators, (6) photochemical and photobiological conversion systems, (7) direct production of fuels and chemicals from concentrated sunlight, and active solar cooling. A few, such as space power satellites, had been abandoned as uneconomic or technically unfeasible.

The federal solar research and development program was marked, in its ten-year history, by a sharp series of twists and turns. In the early years it grew at a phenomenal rate, beginning at $14 million in 1974 and rising to $294 million in 1977, virtually tripling in size every year (Table 1). From 1978 to 1981 the program remained relatively stable in funding at the $400-$600 million level, but in late FY 1981 it began a nosedive that threatened to be as precipitous as its earlier climb. First $95 million was cut from operating funds in the middle of FY 1981; then a reduction of 67 percent in funding was proposed for the next fiscal year (1982); finally, only $72.3 million was requested for all solar operations for FY 1983. Congress agreed to a much gentler drop-off in federal funding, and the program stabilized at the $175-$200 million level.

Table 1
Solar Program Funding, FY 1975-1984
($ in millions)

	Solar Heating/ Cooling	Solar Thermal	Photo-voltaics	Wind	Ocean Systems	Biomass/ Alcohol Fuels	Other	Total
ERDA								
1975	13.6	13.2	5.2	5.7	1.9	0.9	1.4	41.9
1976	43.7	20.6	21.6	14.4	6.1	5.1	3.3	114.9
1977	95.3	79.1	59.7	21.9	14.2	12.7	11.5	294.4
DOE								
1978	113.1	114.7	76.2	35.1	31.2	21.2	9.0	400.5
1979	101.0	109.3	118.8	59.6	41.1	42.4	13.3	485.5
1980	110.3	135.0	157.0	63.4	40.0	56.0	16.0	577.7
1981								
Carter[a]	77.1	154.7	160.2	85.8	39.0	69.8	25.2	578.6
Reagan[b]	68.6	120.0	133.2	54.2	34.6	45.2	27.0	482.8
1982[c]	22.1	56.0	74.0	34.4	20.8	30.5	30.5	268.2
1983[c]	11.7	49.5	58.0	31.4	10.5	21.0	19.9	201.9
1984[c]	17.2	41.5	51.2	26.5	7.0	29.0	6.3	178.7

Notes: [a]Reflects an across-the-board reduction in the overall solar budget made by the Congress.

[b]Actual funding after the first Reagan budget submittal.

[c]Total Reagan administration requests for these years were $195.0, $72.3, $86.9 million.

The program also experienced an inordinate number of organizational changes. In 1974 it was being carried out by the National Science Foundation (NSF) and the National Aeronautics and Space Administration (NASA); in 1975 the responsibilities were transferred to a new Energy Research and Development Administration (ERDA); in 1977 the Department of Energy (DOE) was created and took over the solar R&D program; in 1982 the Reagan administration proposed turning over all of the DOE research and development programs to the Department of Commerce (DOC).

The philosophy guiding the programs underwent equally radical changes. The mandate from Congress in the 1970s was to demonstrate and commercialize solar technologies as rapidly as possible. Thus the "Solar Heating and Cooling Demonstration Act of 1974" (PL 93-409) called for "the demonstration within a three-year period of the practical use of solar heating technology" and "the development and demonstration within a five-year period of the practical use of combined heating and cooling technology" (CRS, 1978). The DOE Organization Act of 1977 (PL 95-91) required the new department to "place major emphasis on the development and commercialization of solar, geothermal, recycling and other technologies utilizing renewable energy resources" (HS&T, 1978: 4). And in the Wind Energy Systems Act of 1980 (PL 96-345) Congress declared that:

> It is the proper and appropriate role for the Federal Government to undertake research and development, to participate in demonstration programs for wind energy systems and assist private industry . . . in hastening the widespread utilization of such systems (H.R. 96-1217; 1980: 2).

The Reagan administration, on the other hand, took a diametrically opposite approach. It proposed in 1981 to:

> Shift the focus of the DOE's solar activities away from costly near-term, development, demonstration and commercialization efforts, and into longer-range research and development . . . It will be the free marketplace - and not government - which will be expected to support the commercial introduction of new and alternative technologies into the marketplace (Reagan, 1981).

Underlying these changes is the question of the proper federal role in the development and commercialization of technologies that are not so far from technical feasibility that they can be safely tucked away in national laboratories or university research centers, but are not mature enough to be ready for commercial application. Four different approaches to this problem characterized American solar energy policy over the past decade. These I have labeled the congressional approach, the ERDA approach, the Carter administration approach, and the Reagan administration approach. Each was somewhat different from the other, each had some severe drawbacks and problems, and each contributed something of value to the development of solar energy technology. The remainder of this paper presents an analysis of these four approaches and some conclusions about the effects of technology, policy and ideology on the outcomes they produced.

The Congressional Approach: The Solar Heating and Cooling Demonstration Act of 1974 (PL 93-409)

The Solar Heating and Cooling Demonstration Act of 1974 was the first piece of direct solar legislation passed by Congress and in many ways was a prototype for those that followed. It had its origins in two days of hearings before the Energy Subcommittee of the House Science and Astronautics Committee in June 1973 (HS&A, 1973a) and three more days of testimony in November 1973 (HS&A, 1973b) on the first draft of a bill prepared by committee staff. Those five days of hearings included statements by over 50 individuals representing six government agencies, as well as researchers and equipment manufacturers, conventional HVAC (heating, ventilation and air conditioning) equipment manufacturers, environmental groups, home builders, architects, engineers, and members of Congress. All recommended that the government take steps to promote solar heating and cooling of buildings, although there was some disagreement about the desirability of the approach proposed in the Heating and Cooling Demonstration Act.

The final version of the act, according to a Congressional Research Service history of the legislation, established "a five year, $60 million program to provide for a major demonstration for solar heating technology in residential and commercial buildings by 1977 and for the development and demonstration of the technology for combined solar heating and cooling in residential and commercial buildings by 1979" (CRS, 1978: 1).

This well-documented, carefully considered initiative did not, unfortunately, produce the results that were expected. The program took longer than anyone anticipated. In 1981, six years after the law was passed, DOE was still funding the federal buildings portion of the commercial demonstration program, and the Department of Housing and Urban Development (HUD) was in the middle of an extensive repair program on the systems it had put in the field. Only 136 of the 287 DOE commercial demonstration projects were fully operational (DOE, 1981) and about 10 percent of the HUD projects financed in 1978 and 1979 were still under construction (BE&C, 1983: Vol. 5,6).

The program cost more than anyone anticipated. Between 1975 and 1981 DOE and HUD spent close to $550 million (see Tables 1 and 2) on solar heating and cooling R&D. About $230 million of that was budgeted directly for the demonstration program, although only $95 million went to direct support of the grantees (DOE, 1981, 3-8; BE&C, 1983: S-4). The program ultimately funded 287 commercial demonstration projects and 497 residential demonstration projects (with an additional 446 awards granted for design only) representing 10,098 living units and 1,255 solar energy systems (BE&C, 1983: S-3).

The average cost of systems in the demonstration program turned out to be much higher than anyone anticipated. At the House Science and Astronautics Committee hearings in 1973 several witnesses confidently predicted that solar heating and cooling systems would cost between $4 and $8 per square foot (HS&A, 1973a; HS&A, 1973b). In fact, in the DOE commercial demonstration program, the system costs ranged from a high of $77/ft^2 in the first cycle to a low of $38/ft^2 for a hot-water only initiative aimed at hotels and motels in 1978 (DOE, 1981: Figures 5-3 and 5-5). HUD reported systems costs in the range of $25-$40 per square foot in its residential demonstration.

Such huge cost increases can be partially attributed, in the DOE program, to government over-specification, unnecessary technical complexity, and gold-plating (GAO, 1980: 5-17). But in the HUD program, where grants were simply given to developers to purchase solar equipment of their own choosing from any manufacturer, such an excuse is not available. Moreover, even today, when a $500 million/year industry exists, systems costs for active hot water and space average more than $40 per square foot (SEIR, 1983: 178). Thus the real problem lay in the underestimation of systems costs by witnesses at the hearings.

Even more distressing than the high costs and lengthy implementation time for the program was the poor performance of the systems. A General Accounting Office investigative team in 1979 found that 46 of the 91 HUD-financed residential units it inspected, or 52 percent, were "either not working or were experiencing problems while in operation" (GAO, 1979: 13). The management support contractor for the HUD Demonstration program (BE&C Engineers, Inc., a Boeing subsidiary) reported in June 1983 that:

> An inordinately high percentage of the total number of solar systems in the demonstration program required significant repair, replacement or removal actions due to serious operating problems. Of a total of 1,255 systems in the program, 599 (48 %) have been the subject of one or more System Operating Problem Reports. All of these systems required at least some technical assistance; most required major repair . . . Collectively these systems depict a horrendous consumer problem (BE&C, 1983: S-5).

The problems occurred mostly in active space heating systems and involved "hazardous installations, corrosion, collector degradation, thermal transfer and storage losses, and control deficiencies" (BE&C, 1983: 63-64).

From a technical point of view, the Heating and Cooling Demonstration Act achieved close to the opposite of its goals. Systems for combined solar heating and cooling were shown to be uneconomic due to the high cost and relative inefficiency of the cooling equipment (this possible outcome was predicted by several conventional cooling equipment manufacturers at the November 1973 hearings). Active space heating equipment was shown to be marginally economic, unreliable and relatively inefficient. Only passive systems turned out to be clear winners - cost-effective and high performance (Table 2). Passive systems, with costs generally less than half that of active systems ($10-$15/ft^2), deliver 50 to 100 percent more energy per square foot of collector aperture.

Table 2
Comparison of Passive and Active Solar Systems, U.S. Department of Housing and Urban Development Demonstration Program, 1983

Mean utility energy consumption of 130 solar homes and 69 comparable dwelling units	
	Btu's/ft^2/degree day
Active solar homes (N=86)	8.72
Passive solar homes (N=69)	6.16
Comparable units (N=69)	13.10
Percentage of incident solar energy delivered to load, instrumented homes	
	Percent (average)
Water heating (N=15)	26
Passive space heating (N=8)	36
Active space heating (N=6)	20
Active space heating (all units monitored by DOE)	11

Sources: Derived from BE&C Engineers, "Final Report of the Management Support Contractor for the Residential Solar Heating Demonstration," 1983, pp. 26 and 43-44; and Frederick H. Morse, presentation at Building Equipment Workshop, U.S. Department of Energy, Airlie, Virginia, December 5, 1983.

To balance this negative picture of the demonstration program, some positive factors must be pointed out. When the program began there was no solar industry to speak of - just a handful of researchers and collector manufacturers. By 1980 there were 133 firms engaged in the manufacture of solar collectors; 161 involved in installation; 195 in systems design and consulting; and 239 in wholesale and 182 in retail distribution (SEIR, 1983: 178). The demonstration program provided a valuable early market for these firms when they were just starting up, before federal tax credits were enacted in late 1978. The program also developed standards and testing procedures for the industry that helped assure the quality of the equipment produced in later years. The program identified many of the early problems with active solar equipment and provided manufacturers and installers with valuable guidance as to how to correct those problems. The program played a key role in developing the infrastructure by which the more cost-effective solar water-heating systems have since been distributed and installed. It also provided a documented record of the costs and performance of systems (much of which has, unfortunately, not yet been published). The reasons for the program's failure to meet congressional expectations are well-summarized in an internal DOE memorandum dated March 18, 1980:

> The Solar Heating and Cooling Demonstration Act was passed by Congress in 1974 when the solar industry was virtually nonexistent. The components available for testing were experimental and mostly unproven; the large scale systems assembled with these components were prototypes . . . Given the state of the art in 1974 and the state of the marketplace . . ., the goals of the Act were premature and unrealistic . . . Most of the early solar demonstration projects were designed by architects and engineers with little, if any, solar experience. The solar equipment was the first of its kind and the manufacturers had no track record of reliability or durability. Production facilities were very limited . . . (Morse, 1980).

Another analyst expressed it this way:

> Demand-oriented commercialization strategies such as market-oriented demonstrations and financial incentives to potential purchasers assume that the technology works and works well, and that key elements of the technology delivery system are in place and functioning well. In the case of solar heating, these assumptions did not hold. A strategy more appropriate to a competitive or mature industry was applied to an emerging one (Roessner, 1984: 243).

The ERDA Approach: Implementation of the Recommendations of the NSF/NASA (1972) and AEC (1973) Panels on Solar Energy

If the Solar Heating and Cooling Demonstration Act of 1974 represented an attempt to commercialize a technology before the necessary systems development and industrial infrastructure building activities had been carried out, two reports by panels convened by the nation's leading scientific and technical agencies avoided that mistake. The NSF/NASA *Assessment of Solar Energy as a National Resource* (NSF/NASA, 1972) and the report of Subpanel IX - "Solar and Other Energy Sources" - to the Atomic Energy Commission's study of U.S. energy R&D priorities, *The Nation's Energy Future* (HS&A, 1974: 170-361) both recommended significant long-term (10 years) research and development programs for a variety of solar technologies. Those technologies included solar heating and cooling, solar thermal energy, wind energy, bioconversion, ocean thermal energy systems, and photovoltaics - the very ones that constituted the ERDA and the DOE R&D programs in later years. Funding levels recommended for the six technologies by the two reports are given in Table 3.

To a large degree these two reports defined the ERDA program as it evolved from 1975-1977. (Compare the technology programs and funding levels proposed by the Subpanel IX report with the evolution of the ERDA program.) This is not surprising given the fact that many of the ERDA program managers who implemented the programs were drawn from the panels that wrote the report. The basic strategy of the program, as recommended by the two panels, was technology development and demonstration, emphasizing the construction of pilot and demonstration plants.

The ERDA program took into account the need for time-consuming research and development to precede demonstrations. In contrast with the approach of the Solar Heating and Cooling Demonstration Act, the Subpanel IX Heating and Cooling program would have first worked on climate and insolation studies, solar design guides and data manuals, component and subsystem development, testing and rating standards, and equipment performance system studies before constructing a modest number of prototype solar buildings. The subpanel advocated the construction of only four single-family residences, two multifamily residences, three commercial buildings, two mobile homes, and four agricultural buildings (in different regions of the country). A demonstration program consisting of the design and construction of several "small solar communities" would be carried out last. It was estimated that these activities would lead to the commercial availability of solar heating and cooling technologies by 1979 (HS&A, 1974: 213-214).

Table 8
AEC Subpanel IX
Recommended Solar Energy Program Resources
(Millions of $)

Program	1975	1976	1977	1978	1979	Totals	NSF/NASA 10-yr. Program
Heating/Cooling							
a. Accelerated	17.8	35.7	50.7	57.5	42.6	204.3	100
b. Minimum	8.1	20.2	22.0	19.0	17.8	86.9	
Solar Thermal							
a. Accelerated	31.1	51.1	59.0	60.0	74.0	275.2	1130
b. Minimum	15.8	16.8	31.5	43.0	38.0	145.1	
Wind Energy							
a. Accelerated	8.5	17.3	23.5	33.9	23.0	106.2	610
b. Minimum	3.8	5.4	7.1	6.9	3.7	26.9	
Bioconversion							
a. Accelerated	12.1	17.3	23.7	30.1	40.6	123.8	370
b. Minimum	8.5	10.5	10.5	10.5	12.5	52.5	
Ocean Thermal							
a. Accelerated	8.6	19.2	16.4	28.8	26.5	99.5	530
b. Minimum	4.0	4.5	7.6	12.2	13.0	41.3	
Photovoltaic							
a. Accelerated	28.3	48.3	63.9	54.1	53.1	247.7	780
b. Minimum	10.3	10.3	11.0	13.2	12.4	57.2	
Totals							
a. Accelerated	106.4	188.9	237.2	264.4	259.8	1056.7	3520
b. Minimum	50.5	67.5	89.7	104.8	97.4	409.9	

Source: U.S. House Committee on Science and Astronautics, Hearings on the Solar Research, Development and Demonstration Act of 1974, July 30 - August 2, 1974.

Despite the willingness of the ERDA program to take more time and pay respect to the development and testing requirements of the innovation process, there were certain similarities with the Solar Heating and Cooling Demonstration Act. Both tended to be hardware and technology oriented, and thus both missed simpler, lower-cost and more readily implementable options such as passive solar design and solar greenhouses. Both tended to underestimate the time and resources required to meet the objectives established. The Subpanel IX wind program, for example, expected to design and construct several multi-megawatt units and a 10 megawatt wind farm by 1979 at a total budget of $106 million (HS&A, 1974; 241-242; 253-254). As it turned out, the federal wind program spent $290 million from 1974-1981, but only built four multi-megawatt machines and was still one generation away from a cost-competitive model at the close of that effort. The Subpanel IX solar thermal program claimed that it could build a 10 MWe pilot plant by 1978 for $14 million (HS&A, 1974: 223-237). In fact the Solar One 10 MWe plant at Barstow, California, cost $140 million and was not completed until 1982. (In fairness, the NSF/NASA solar thermal panel estimated a ten-year time period and a cost of $100 million to construct the first 10 MWe central receiver pilot plant) (NSF/NASA, 1972: 310-311). The OTEC (ocean thermal energy conversion) subpanel estimated that it could construct two 5-10 MWe pilot plants by 1979 at a total cost of $40 million (HS&A, 1974: 299). In fact, it cost more than $40 million to outfit a 1 MWe test ship in 1980, and the first pilot plant is today only in preliminary design and will cost more than $200 million for a 40 MWe unit, if private financing can be found to build it. (All actual expenditures are from DOE program documents.)

The most telling criticism of the ERDA program, which was articulated several times in the 1978-1980 time period, was that it was "making solar after the nuclear model" (Metz, 1978). This criticism in fact consisted of two separate charges. First, a number of studies pointed out that the ERDA program overemphasized electricity-producing technologies from central station power plant applications at the expense of decentralized thermal and fuel applications. For example two reporters for *Science* characterized the program as follows:

> Despite the diffuse nature of the resource, the research program has emphasized large central stations to produce solar electricity in some distant future and has largely ignored small solar devices for producing on-site power - an approach one critic describes as 'creating solar technologies in the image of nuclear power.' The program contains virtually no significant projects to develop solar energy as a source of fuels and only modest efforts to exploit it as a source of heat. The massive engineering projects designed by aerospace companies which dominate much of the program seem to be intended for the existing utility industry - rather than individuals or communities - as the ultimate consumer of solar energy equipment (Metz, 1978: 1).

Critics pointed out that electricity accounts for less than one-tenth of the nation's end-use energy and less than one-third of its annual fuel consumption. Furthermore there are many attractive applications for direct solar heating that displace the scarcer liquid and gaseous fossil fuels. The NSF/NASA report, by contrast, proposed that 87 percent of the solar R&D budgets be devoted to solar electric technologies. The Subpanel IX report proposed 69 percent; the ERDA 1977 budget had 62 percent of its funds devoted to solar electricity.

This is a valid criticism, but certain mitigating factors must be kept in mind. First, some technologies simply require more research and development dollars than others. Building a solar thermal central receiver power plant is inherently a $100 million endeavor. Conducting research to improve the efficiency of flatplate solar collectors, on the other hand, can be done effectively for 10 percent of that cost.

Second, some solar electric technologies can in fact serve more than the utility market. The central receiver technology produces steam to drive a turbine, but that same steam could also be used to power an industrial process. The industrial process heat market, however, is much more difficult to penetrate than the electric utility market. Industrial processes are quite varied in their operating characteristics, whereas utilities generally require steam at relatively uniform temperatures and pressures. Industrial managers require paybacks on new energy technologies of three to four years; utilities are constrained by regulators to examine the benefits of a new technology over its entire lifetime - fifteen to thirty years. Thus, it makes sense, as the ERDA program did, to aim the first applications of this technology at the utility market and wait until mass production has reduced the cost of heliostats sufficiently to make systems attractive in the more demanding industrial process heat market.

Third, for many renewable energy technologies (as for conventional technologies), there are apparent economies of scale, including reductions in operations and maintenance costs per unit of energy produced, that lead toward larger systems. This seems especially true for OTEC systems and solar thermal central receivers.

The second part of the "solar in the image of nuclear energy" criticism, levelled at the program by small businessmen and appropriate technologists, was that the vast proponderance of funding went to large corporations and national laboratories rather than small businesses and individual inventors. The solar thermal program, for example, spent millions of dollars at Sandia National Laboratories, McDonnell Douglas Corporation, Martin Marietta, Rockwell International, and Boeing. The wind program spent like amounts at General Electric, Boeing, Hamiltion Standard, Rockwell International, and NASA Lewis Space Center. The OTEC program devoted the lion's share of its budget to contracts with General Electric, TRW, Lockheed, and Argonne National Laboratories.

Meanwhile, dozens of small businessmen were unable to get their concepts evaluated by the solar program, and hundreds of backyard tinkerers could not find support for low-cost, unconventional solar approaches. In the words of one of the more extreme critics of the ERDA program:

> By May of 1975 the Solar Division of ERDA had received more than 5,000 inquiries and the National Bureau of Standards more than 600 letters requesting evaluations of solar inventions . . . Such a response left no question that small-scale America was ready to move on solar power. But the federal government was not all ready for small-scale America. Sometimes politely and sometimes brusquely, government officials denied the overwhelming majority of requests for support by small solar firms and inventors. Grant applications were returned scarcely opened . . . (Reece, 1979: 70-71).

This despite the fact that, according to the House Committee on Small Business, "for the period between 1953 and 1973 . . . small firms produced about four times as many innovations per research and development dollar as medium sized firms, and about 24 times as many as the largest firms" (Reece, 1979: 83).

The difficulty of access to the solar program for small businessmen, individuals, community groups, and other non-corporate entities was a persistent and well-documented problem (HCSB, 1979). Its cause lay in part in the origins of ERDA in AEC and NASA. Neither of these agencies had a great deal of experience dealing with individual inventors, grass-roots organizations, or small businesses. Both had procurement practices that favored those who have Washington offices, read the *Commerce Business Daily* daily, have staff available to prepare lengthy proposals, and can afford to wait a year for a competition to be completed and an award to be announced, not to mention two years for final payment. Naturally such practices tend to favor large corporations. In addition, the ERDA program managers, largely drawn from those two agencies and NSF, were accustomed to implementing well-defined programs, but not to funding a variety of interesting but tangential innovations. Furthermore, with budgets tripling every year and staffing levels remaining constant and a procurement office that was often better at losing forms than processing them, the program managers were swamped (Metz, 1978: 2). It was far easier to award a $5 million contract to General Electric than to attempt to process one hundred $50,000 procurement requests to individual inventors and small businessmen.

Two special programs were set up, in part, to deal with this situation. The Energy-Related Inventions Program provided for review of a potential invention by the National Bureau of Standards (NBS), and upon successful completion of that review, development funding from ERDA if an agreement could be negotiated between the inventor and the agency. More than eighteen thousand evaluation requests were received by NBS between 1974 and 1982; nine thousand were actually evaluated, 219 recommended to ERDA or DOE for funding; 165 inventions were actually supported. In addition, an Appropriate Technology Small Grants program was set up to make awards of up to $50,000 to individuals, small businesses, community groups, and state and local agencies for low technology development, demonstration and education projects involving conservation and renewable energy resources. More than 45,000 proposals were received, and nearly 2,400 were funded in the four year life of this program (1978-1981). (All figures are from DOE program documents.) In addition Congress imposed a number of small business set-aside requirements on the ERDA/DOE solar program in the 1977-1980 time frame. None of the above measures, however, fully integrated "small scale America" into the solar program. In fact, the two special programs discussed above were run, not out of the solar program but out of the conservation office. All of the initiatives left far more customers unsatisfied than they helped, as can be seen from the ratio of awards made to proposals received.

In fairness to the solar program, it should be pointed out that some tasks, such as designing and building a 3 MWe wind machine, are best carried out by large engineering firms like Boeing and General Electric. These tasks involve engineering development and construction which is the strong point of such companies. In addition, it is the large corporations that have the capital and the mass production capability to manufacture solar equipment cheaply. They are likely to become involved in solar energy no matter what the government does. The participation of large corporations in the decentralized solar applications are examples: Exxon, GE, Grumman, Revere, PPG, Owens Illinois in flat plate thermal collectors; ARCO, Sohio, Mobil, and Amoco in photovoltaics. The solar advocates of the 1970s who wanted to keep solar energy as the province of the small businessman and individual entrepreneur were either being extremely naive about the way the world works or overly optimistic about the ability of technology to force social change.

The Carter Administration Approach: The Solar Domestic Policy Review of 1978

The Nixon and Ford administrations were largely indifferent to solar energy. In December 1973 the AEC report, *The Nation's Energy Future,* recommended a five-year solar budget of about $40 million a year, which was about half of the "minimum viable" program defined by Subpanel IX. President Ford's funding requests for solar for FY 1976 and FY 1977 were each substantially increased by Congress (from $40 million to $118 million in FY 1976 and from $160 million to $290 million in FY 1977).

The election of President Carter in 1976 raised the hopes of solar energy advocates. Carter was sworn in on a solar-heated reviewing stand, promised to "vigorously expand the use of nonconventional energy sources" in his National Energy Plan, and endorsed for the first time in his National Energy Act a set of federal solar tax credits which Congress had long been advocating.

Disappointment quickly set in, however. The first Carter budget request for solar energy for FY 1978 was identical in dollar amount to the budget previously submitted to Congress by President Ford. The FY 1979 request actually called for a reduction from the FY 1978 appropriation (Metz, 1978: 7). Carter's nominee for Secretary of Energy was James Schlesinger, former head of the Atomic Energy Commission and the Department of Defense, a man who hardly gave confidence that the perceived pro-nuclear policies of ERDA would be changed. In addition, when the solar tax credits became bogged down in the wrangling over natural gas pricing and other controversial provisions of the NEA, the President refused to allow them to be passed separately, in part in order to force solar advocates to support his entire package. The one-year delay in enacting solar tax credits significantly hurt the solar industry, since consumers withheld purchases of solar equipment awaiting a better deal after passage of the tax credits.

Frustrated by their inability to get a meaningful commitment to solar energy out of the Carter administration, solar energy advocates sponsored a national Sun Day on May 3, 1978, to mobilize public sentiment for renewable energy. President Carter, now on the defensive, announced a $100 million supplemental budget request for the DOE solar programs for FY 1979 and the formation of a Domestic Policy Review of Solar Energy. The purpose of this multi-agency review was, according to the Response Memorandum which resulted from it, to provide the President with:

1. a thorough review of the current federal solar programs to determine whether they, taken as a whole, represent an optimal program for bringing solar technologies into widespread commercial use on an accelerated timetable;

2. a sound analysis of the contribution which solar energy can make to U.S. and international energy demand, both in the short and the longer term;

3. recommendations for an overall solar strategy to pull together federal, state and private efforts to accelerate the use of solar technologies (DOE, 1979: 1).

The review was carried out by an interagency task force under the chairmanship of the Secretary of Energy which included more than 100 officials representing over 30 agencies working for close to six months.

This was the first attempt to conduct a comprehensive analysis of the policy options available to the federal government for promoting solar energy. It succeeded admirably in its first objective, identifying a surprising number of existing programs spread throughout the federal government which were supporting, or could be easily modified to support solar energy (DOE, 1979: 7-20). It also identified a large number of simple initiatives which could make those programs more helpful in promoting solar development and commercialization (DOE, 1979: 39).

The Domestic Policy Review (DPR) also succeeded quite well in its second objective, defining the ultimate solar potential for the year 2000. This was a subject of considerable controversy in 1978, with estimates ranging from a low of 0.1 percent of total nation energy consumption (CONAES Low Solar case) to 37 percent (Armory Lovins) (see Table 4) (DOE, 1978: 42). It had also become something of a litmus test for solar advocates of President Carter's devotion to their cause. The DPR addressed this issue by forming a panel of experts, ranging from solar skeptics to solar enthusiasts, and having them argue with each other for hours until a consensus was reached on the potential contribution of each solar technology to the national total.

The result was a Base Case, which assumed no new solar initiatives, in which solar energy contributed 10 to 13 percent of total national energy consumption in the year 2000 (95-114 quads); a Maximum Practical Case, which represented the task force's best estimate of the largest possible solar contribution short of all-out national mobilization (18.5 quads); and a Technical Limits Case of 28.5 quads (DOE, 1979: 24-25). Probably the greatest single accomplishment of the Solar DPR was in finding a solar contribution and a solar goal (see below) for the year 2000 that Denis Hayes of the Solar Lobby and Energy Department Secretary James Schlesinger could agree to recommend to the President.

Table 4
Summary of Projected Solar Impacts
Primary Quads Displaced, Year 2000

Scenario:	Total	Solar[a]	Percent
Mitre SPURR NEP	115	6.0	8.0
Mitre SPURR RTS	113	5.0	4.4
SRI-Reference Case	145	6.0	4.1
SRI-Solar Emphasis Case	148	15.0	10.1
SRI-Low Demand Case	89	7.0	7.9
Solar Working Group -			
with Behavioral Lag	115	8.1	7.0
without Behavioral Lag	115	8.8	7.7
MOPPS	117	2.8	2.4
IERPS	132	5.8	4.3
Lovins[b]	108	40.0	37.0
NSF/NASA	177	11.8	6.6
Project Indep-B.A.U.	180	10.8	6.0
- Accelerated Solar	180	38.8	21.3
CONAES - High Solar	146	14.0	9.6
- Low Solar	146	0.1	0.1
ERDA-49	150	10.0	6.7
CEQ-April 1978[c]	100	24.5	24.5

Notes: [a]Solar components exclude hydropower. Some studies also exclude the current biomass consumption.

[b]Includes all soft technologies.

[c]Mid-range of CEQ estimates for each technology.

As to setting a solar strategy, which was the third objective, the DPR represented at a best a modest step in the right direction. The Carter administration's weak implementation of the DPR recommendations turned that modest step into total failure.

The DPR recommended that President Carter set a goal of 20 percent solar energy use nationwide by the year 2000. That percentage represented the smallest value the solar advocates would accept, and the largest that could be supported by analysis (the Maximum Practical Case) and the conservatives in the Carter administration (DOE, 1979: 43-45). The DPR also recommended a modest set of targeted initiatives to speed the development of promising near-term solar technologies. These included a Solar Bank, a 20 percent tax credit for industrial users of solar equipment, a 20 percent passive solar tax credit for home builders, an expanded R&D program aimed at nearer-term technologies to displace oil and natural gas and greater use of renewable energy equipment by the federal agencies and the federal power marketing administrations, such as TVA and the Bonneville Power Authority. (Actually, this was the second of three options presented to the President. The first, consisting of the administrative changes mentioned above, was widely perceived as a "do-nothing" option. The third, an all-out initiative which was put together at the last minute by the Solar Lobby, was never seriously considered by the Carter administration because of its budget implications.)

The problem with this "strategy" was that the initiatives proposed were not sufficient to reach the solar goal of 20 percent. Taken together with the Base Case, they at best added up to a solar penetration of 15 percent by the year 2000 (DOE, 1979: 39). Thus the Carter administration was placed in the awkward position of setting a goal without proposing a program to achieve it. This problem was further compounded by the failure to move swiftly and enthusiastically to implement the DPR recommendations. The Response Memorandum languished on the President's desk for six months, from December 1978 to June 1979. By that time events in Iran had triggered a second major oil shortage and a series of price increases, and President Carter had already announced a much larger energy initiative - the decontrol of oil prices and the imposition of a crude oil windfall profits tax. He had angered solar advocates by tying several of the more popular initiatives from the Solar DPR to the oil decontrol measure, and he further angered them in July 1979 by announcing a proposal for a $100 billion Energy Security Corporation to underwrite the production of synthetic fuels, an initiative that dwarfed the $3 billion he had reluctantly agreed to spend for solar development.

To make matters worse, a year after the DPR the General Accounting Office issued a report which found that the Carter administration still had no plans for achieving the 20 percent solar goal and had failed to implement several of the administrative measures announced in June 1979, including formation of a solar policy coordinating subcommittee (GAO, 1980: 1). An Office of Technology Assessment (OTA) study found the solar programs to be understaffed, lacking in direction, and suffering from "a pervasive belief within and outside DOE that senior DOE management does not really care" about them (OTA, 1980: 4). Hearings before the House Science and Technology Committee in June 1980 established that DOE's projected out-year solar budgets would not be sufficient to meet the 20 percent goal, and that the department was planning a significant expansion of its coal and nuclear programs while holding conservation and solar essentially constant (HS&T, 1980: 1-2).

The Solar DPR thus completely failed to satisfy President Carter's liberal critics while it convinced conservatives he had become a captive of the "solar cultists." The DPR did succeed in increasing the resources proposed for existing solar programs in President Carter's FY 1980 budget request and in redirecting some of the research toward technologies that would directly displace oil and natural gas. The industrial tax credits became law and have played a major role in stimulating a variety of third party-financed renewable energy projects. However, those tax credits were reduced in size from 20 to 15 percent and in duration from 10 years to 5 years. The renewable energy industry now faces the unpleasant prospect of losing the credits in 1986, before it has had a chance to establish a solid industrial base. The passive tax credits were never enacted. The Solar Bank has been rendered inoperative by the Reagan administration's refusal to staff it or request funds for it. Only an annual congressional appropriation of about 5 percent of the amount originally envisaged - $500 million/year - and a successful law suit have kept the Solar Bank from completely disappearing. No one talks about a solar goal for the year 2000 any more.

The Reagan Administration Approach: The Free Market Takes Over With a Vengeance

Unlike the Carter administration, which was sometimes characterized as suffering from "paralysis through analysis," the Reagan team required little time or thought to implement a radically revised energy policy. The new director of OMB, David Stockman, had been a leading advocate of a free-market approach to energy in his tenure in Congress. In a highly influential article for *Public Interest* in 1978, Stockman had argued that energy should be treated like any other commodity and allowed to equilibrate supply with demand at an appropriate market clearing price. Any government energy policy constituted an unjustified interference in the workings of the marketplace. The new Secretary of Energy, James Edwards, was a strong advocate of nuclear power who did not like regulation and had no expertise or interest in conservation or solar energy.

Between them, they formulated an energy research and development policy which amounted to unfettered government support for nuclear energy and the free market for everything else. The government was to take all actions necessary to assure the timely development and deployment of nuclear power. For all other energy resources, the only appropriate federal role was to carry out "long-term high-risk high payoff" basic and applied research. Since the solar programs were oriented toward development and demonstration, they would, of necessity, receive major cuts. This was not, however, due to any malice on the Reagan administration's part - so officials assured Congress - but rather to the near-term status of the solar technologies and the need to cut the federal budget. The administration expressed its conviction that the marketplace would do a better job of developing and commercializing solar technologies than the government could possibly do. Doubters were assured that the higher energy prices that would result from the decontrol of oil and the expanded business opportunities that would accompany the economic recovery triggered by the administration's policies would more than compensate for the loss in federal funds.

The Reagan administration's promises quickly failed to materialize. Economic recovery turned into the worst recession since World War II. Oil prices began to drop, rather than rise, reducing the private incentive for investment in new energy technologies. Yet the administration clung to its solar policies and proposed for FY 1983 even more draconian cuts than in its FY 1982 budget request (a 67 percent cut in FY 1982; a 73 percent cut in FY 1983). Even the solar business tax credits, which helped to keep many renewable energy businesses alive, were proposed for rescission in the FY 1983 budget request (while similar tax breaks for oil and natural gas drilling were left untouched).

At this point Congress, which had agreed to much of the Reagan administration's energy budget reduction proposals in 1981, refused to comply. The FY 1983 budget requests for conservation, solar and fossil energy were rejected, as was the President's plan to dismantle the Department of Energy and turn its functions over to the Department of Commerce. The Secretary of Energy and his Assistant Secretary for Conservation and Renewable Energy were attacked at hearings by conservative Republicans for their failure to take non-nuclear energy, and especially energy conservation, seriously, and to present a balanced energy budget. (See, for example, the Statement of Hon. Vin Weber, a Member of Congress from the State of Minnesota in HS&T, 1982: 10-11.) The Reagan administration thus wound up alienating many of its supporters by acting too hastily, just as the Carter administration alienated many of its supporters by acting too cautiously.

The effect on the solar programs of the Reagan administration's budget cuts and its reversal in federal policy was quite severe. Subprograms which had been progressing toward well-defined goals were abruptly redirected or terminated. Research teams in industry and at the national laboratories were disbanded or reduced to skeletal staff. Experienced program managers at DOE were removed and replaced by personnel with limited background or skills in the relevant research areas. Some of the more fortunate subprograms were simply reduced in size and scope.

More seriously, by attempting to eliminate all technology development and demonstration activities, the Reagan administration rendered ineffective the very vehicle by which technology transfer to private industry could have taken place. Industry and national laboratory officials have consistently testified that to commercialize a new energy technology, the private sector must see a working model of reasonable size from which it can make projections as to the likely cost and performance of commercial systems. Without such a demonstration, the cost of engineering development, the risk of failure, and the time to realize a payback are all too great to make the technology development investment appear worthwhile to senior corporate management. By refusing to allow the solar programs to build anything bigger than a bench model, the Reagan administration in effect made it likely that much of the solar research done by DOE would never be transferred to industry.

The commercialization of the solar technologies since 1981 reflects the consequences of that policy decision. Those technologies which were already in the commercial marketplace in 1981 - such as flat plate collectors, low-head hydro and small wind machines - have achieved somewhat tenuous market niches, based largely on the federal solar tax credits that terminate in 1986. Those technologies that were still in the development and demonstration stage in 1981 - such as solar thermal parabolic troughs and central receivers and large wind machines - are experiencing great difficulties. Without government-assisted demonstration projects to sustain them through the early prototype phase, they are forced to utilize the third-party project financing mechanism that is more appropriate to commercial technologies. These projects are marginally economic and often fall apart when investors require performance guarantees that the equipment manufacturers are unable to provide. It thus appears that many of the more promising solar technologies developed by the ERDA/DOE solar program will fail to make the transition from technology development to commercialization, in part because of the Reagan administration's decision to demolish the bridges that usually span that gap.

Conclusion

It would be unwise to attempt to draw sweeping conclusions from a limited examination of four initiatives related to one set of technologies. The examples considered, however, illustrate a number of points about the influence of technology, politics and ideology in the formulation of U.S. solar energy policy.

Technology

It takes about 15 years to develop and diffuse a technology from initial concept to final market penetration (HS&A, 1973b: 118, 402). That time span encompasses a number of necessary steps which apparently cannot be circumvented merely by the will of Congress or President. Crash programs to demonstrate and commercialize technology are often unsuccessful. A premature demonstration or commercialization program can waste government resources on technologies that are defective or inappropriate. It can also lead to the neglect of more viable approaches to the same problem. On the other hand, cost-shared demonstrations that follow a carefully planned research and development program can be successful and appear to be necessary in aiding the transfer of technology to industry. Utilities have testified that the Barstow 10 MWe pilot plant and the MOD2 wind turbines, for examples, are crucial to their interest in solar thermal and large wind technologies (HS&T, 1983a: 584). Tax credits, which allow the marketplace to select the technology and the application but guide the market in directions favored by overall national policy, also appear to be effective in aiding the diffusion of new technologies, such as small wind turbines (HS&T, 1983b).

Politics

The orderly and rational technology development and commercialization process that exists in textbooks and theoretical models is considerably rearranged by the political process. Congressional attempts to bypass the full development and testing process for active solar heating and cooling systems led to a rather costly and unsuccessful demonstration program. Similar programs for synthetic fuels, alcohol fuels and nuclear reactors have met similar fates. The Reagan administration's rush to reverse overnight six years of energy policy wound up impeding the technology transfer process it claimed to be attempting to accomplish. A largely symbolic goal, set for political rather than programmatic reasons, wound up embarrassing rather than helping the Carter administration's efforts to support solar energy. Probably the most serious problem posed by politics for technology development is the balancing of a desire to demonstrate a commitment or achieve an early result with the actual time it takes to accomplish significant technological change.

Ideology

Three different ideologies appear to have affected the evolution and operation of the U.S. solar program. One was that of the "solar romantics," who saw solar energy as ushering in an age of "profound and multi-dimensional transformation in the political, economic and social milieu."

> The arrival of the solar age [according to one account of this view] will signal the triumph of democracy, the accommodation to human scale, the achievement of local self-determination, the fulfillment of the virtues of material simplicity and personal self-reliance, and the emergence of a universal commitment to ecological harmony (Byrne and Rich, 1983: 164-165).

The second was that of the free market true believers for whom the marketplace was by definition good and government by definition evil. For them the fact that the oil industry is dominated by an international cartel and a dozen immense multinational corporations; that the electricity and natural gas markets are both state-regulated monopolies; and that conventional energy resources have received tens of billions of dollars in government subsidies are irrelevant to their belief in the existence and virtue of a "free energy market."

Third was the ideology of the technologists. They tended to see problems that were economic, political and social, as well as technological, as ultimately curable by the development of new hardware.

The solar romantics helped the U.S. solar programs by focusing the attention of public officials on a source of energy that had largely been overlooked in the fascination with large, expensive coal and nuclear hardware. But by making the rapid transition to a Solar Age the over-arching goal against which all policies had to be measured, the romantics distorted solar policy making and reduced the credibility of the technology and its backers.

The Reagan administration free marketers' stress on broad market incentives to investment - such as accelerated depreciation - and the role of the marketplace in selecting new technologies were a helpful corrective to previous policies which attempted to push favorite energy systems. Their failure to apply their principles consistently (such as their continued support for the Clinch River Breeder Reactor and tax incentives for oil and gas drilling) and their failure to recognize or deal with market failures and disincentives to the development of new technologies, however, caused them to distort even further the process of federal intervention they so loudly deplored.

Finally, the technologists succeeded in defining a set of challenging scientific and technical objectives for the engineers associated with the solar programs to accomplish. But their preoccupation with exotic hardware caused them to devalue simpler, non-technical approaches and government policies outside the research and development process that might have accomplished the objectives of the federal solar effort in a more timely manner.

In general, the history of the national solar energy program over the past decade illustrates the importance of the interaction between technology, politics and ideology in shaping the outcome of federal research and development policy. Ideology constrains the choices key actors make as to which policies to pursue; politics determines the effectiveness of translating those policies into programs and budgets; and the characteristics of the technology - cost, performance, and state of maturity - limit the degree to which the goals of the programs are actually realized. It is only by taking all three factors into account that we can hope to obtain a full understanding of the government R&D process.

REFERENCES

BE&C (BE&C Engineers, Inc.) (A Boeing Subsidiary). 1983. "Final Report of the Management Support Contractor for the Residential Solar Heating Demonstration." (Unpublished report to the U.S. Department of Housing and Urban Development.)

Byrne, John and Daniel Rich. 1983. "The Solar Energy Transition As A Problem of Political Economy." In Daniel Rich, Jon M. Veigel, Allen M. Barnett, and John Byrne, *The Solar Energy Transition.* Boulder, CO: Westview Press for the American Association for the Advancement of Science (Symposium #74), pp. 163-86.

CRS (Congressional Research Service). 1978. U.S. Congress. *Energy Research and Development Administration. Statutes and Legislative Histories, Volume I. Solar Heating and Cooling Demonstration Act of 1974. Public Law 93-409.* Printed by the Committee on Science and Technology (HS&T), U.S. House of Representatives, April 1978.

DOE (U.S. Department of Energy). 1981. "Solar Heating and Cooling Commercial Demonstration Program Assessment." Unpublished report to the U.S. Department of Energy, Vol. 1 (June 1981).

______________. 1979. *Domestic Policy Review of Solar Energy* TD 22834, February 1979.

______________. 1978. *Solar Energy: A Status Report* DOE/ET-0062 (June 1978).

GAO (General Accounting Office). 1980a. U.S. Comptroller General, Report to Congress, "Federal Demonstrations of Solar Heating and Cooling on Commercial Buildings Have Not Been Very Effective." EMD 80-41. April 15, 1980.

______________. 1980b. U.S. Comptroller General, "The 20-Percent Solar Energy Goal, Is There a Plan to Attain It?" Report to Congress #EMD 80-64, March 31,1980.

______________. 1979. U.S. Comptroller General, Report to Congress, "Federal Demonstrations of Solar Heating and Cooling on Private Residences - Only Limited Success." EMD 79-55. October 9, 1979.

HCSB (House Committee on Small Business). 1979. U.S. Congress. Subcommittee on Energy Environment and Safety. "Role of Government Funding and its Impact on Small Businesses in the Solar Energy Industry." 96th Congress, 1st session. December, 1979.

H.R. 96-1217. 1980. U.S. House of Representatives, Report No. 96-1217, Wind Energy Systems Act of 1980.

HS&A (House Committee on Science and Astronautics). 1974. U.S. Congress. Subcommittee on Energy. Solar Energy Research,

Development and Demonstration Act of 1974. Hearings, 93rd Congress, 2nd session, July 30, August 2, 1974, pp. 170-361.

______________. 1973a. U.S. Congress. Subcommittee on Energy. Solar Energy for Heating and Cooling. Hearings, 93rd Congress, 1st session. June 7 and 12, 1973.

______________. 1973b. U.S. Congress. Subcommittee on Energy. Solar Heating and Cooling Demonstration Act. Hearings, 93rd Congress, 1st session. November 13, 14 and 15, 1973.

HS&T (House Committee on Science and Technology). 1983a. U.S. Congress. Subcommittee on Energy Development and Applications. "Fiscal Year 1984 Department of Energy Authorization (Conservation and Renewable Energy)." Hearings, 98th Congress, 1st session, March 15, 1983.

______________. 1983b. U.S. Congress. Subcommittee on Energy Development and Applications. "Tax Credits for New Energy Technologies." Hearings, 98th Congress, 1st session. July 12, 21, 1983.

______________. 1982. U.S. Congress. Fiscal Year 1983 Department of Energy Budget Review Hearing. 97th Congress, 1st session. March 4, 1982.

______________. 1980. U.S. Congress. Subcommittee on Energy Development Applications. "Oversight: DOE Solar and Conservation Programs." Hearings. 96th Congress, 2nd session. June 12, 1980, pp. 1-2.

______________. 1978. U.S. Congress. Subcommittee on Fossil and Nuclear Energy Research, Development and Demonstration. *Principal Energy Research and Development Legislation Through the 95th Congress,* December 1978.

Metz, William D. and Allan L. Hammond. 1978. *Solar Energy in America.* Boulder, CO: Westview Press for the American Association for the Advancement of Science.

Moore, J. Glen. 1983. Personal communication. Specialist in energy technology, Congressional Research Service (December 15).

Morse, Frederick H. 1983. Presentation at the Building Equipment Workshop, U.S. Department of Energy, Airlie, VA., December 5, 1983.

______________. 1980. Memorandum of March 18 from him as Director, Office of Solar Applications for Buildings (U.S. Department of Energy), to Donald Gestiehr, Office of the Controller: "SA Response to GAO Report on Commercial Solar Demonstration Program."

NSF/NASA (National Science Foundation/National Aeronautics and Space Administration). 1972. NSF/NASA Solar Energy Panel,

An Assessment of Solar Energy as a National Energy Resource. Report to the National Science Foundation, December 1972.

OTA (Office of Technology Assessment). 1980. U.S. Congress. *Conservation and Solar Energy Programs of the Department of Energy: A Critique, 1980.*

Reece, Ray. 1979. *The Sun Betrayed: A Report on the Corporate Seizure of U.S. Solar Energy Development.* Boston, MA: South End Press.

Roessner, J. David. 1984. "Commercializing Solar Technology: The Government Role." *Research Policy* 13: 235-46.

SEIR *(Solar Energy Intelligence Reporter).* May 30, 1983.

Stockman, David. 1978. "The Wrong War? The Case Against a National Energy Policy." *The Public Interest,* May 1978, pp. 1-44.

The Federal Role in Conservation Research and Development

Maxine Savitz

Energy conservation involves activities that reduce energy consumption through improved efficiency. While conservation is pursued to avoid energy waste, true long-term conservation should lead to the production of more goods and services per unit of energy consumed. Over the past ten years, it has been demonstrated that conservation can be obtained without abandoning personal freedoms and mobility or reducing one's standard of living. Moreover, conservation should be viewed as an important resource during the transition to renewable energy.

There already has been a significant shift to energy efficiency. Automobiles made by the big three domestic manufacturers averaged 12 mpg in 1973; today the figure is close to 24 mpg. Residential and commercial buildings in the United States consume 20 percent less energy per square foot than they did ten years ago. The industrial sector has been the most responsive: energy consumption per unit of production in the manufacturing industry fell by 23 percent since 1973. Moreover, many significant improvements in energy efficiency have been accomplished through "housekeeping" retrofits and low-cost measures that did not require high capital investment.

While past savings are impressive, there is still potential and opportunity for greater efficiency in all sectors on the order of 20-40 percent. These savings can be achieved at an equivalent cost of less than $20 a barrel of oil - a very cost-effective alternative to imports at $29 a barrel and new supplies ranging from $40-60 a barrel for synthetic fuels and other alternative fuels. The need to achieve this additional energy conservation should be clear. We currently import approximately 29 percent of our oil needs and pay approximately $65-70 billion a year for this imported oil (EIA, 1985).

If we are to exploit the potential for greater energy conservation, we must recognize and address the special characteristics that make this option quite different from conventional energy supply options. First, energy conservation actions are usually undertaken by millions of individual decision makers rather than a few decision makers. Second, energy conservation technologies are diverse and small in scale so that research and development must be distributed among a variety of relatively small projects, rather than concentrated in multimillion dollar, capital-intensive systems. Finally, the success of energy conservation depends upon penetration of conservation technologies into the market and the signals the market sends. Thus, many of the important R&D questions in energy conservation relate to the issues of market penetration rather than only whether a particular technology will work in a reliable and productive manner.

There is a desire and a need to continue progress toward a more energy efficient economy during the next ten years. The challenge of the 1980s requires a comprehensive look at energy and technology trends. An analysis of the types of mechanisms that have been used in this country, Europe and Japan - both private and public - to introduce new technologies should be conducted in order to determine the most effective means for complementing market signals and ensuring that the U.S. continues to become a more energy efficient society.

Given the need for greater energy efficiency, two major questions need to be addressed: What is the appropriate national strategy to encourage conservation? Within this strategy, what should be the role of federal R&D funding? The analysis which follows is intended to facilitate consideration of these questions. It begins with an examination of the rationale for a federal role in energy conservation research, development and demonstration. With this perspective, the historical development of federal conservation programs is reviewed, the specific accomplishments of these programs are assessed, and the opportunities for future federal R&D investment are identified. Finally, the current challenge to federal conservation R&D is considered and evaluated.

The Rationale for a Federal Role in Conservation Research, Development and Demonstration

While substantial improvements in energy efficiency have been occurring particularly through the application of known technologies, more can be gained through appropriate research and development

(EIA, 1984). Federal support can accelerate this research and development, share in the risk with the private sector, and contribute to the nation's security both in terms of reduced oil vulnerability and enhanced technological capabilities. The federal role must be structured to complement the marketplace. In a market economy, higher energy prices help drive both the use of existing conservation technologies and the R&D necessary to support future technology developments. But markets do not work perfectly, and they are not capable of carrying out all desirable R&D goals. In energy, freely competitive markets simply do not exist for either natural gas or electricity; prices are regulated, generally to the average cost of old and new sources rather than the cost of higher new sources. Thus the current market does not automatically bring forth as large an investment in conservation practices or R&D as is justified economically. Also, the implementation of energy conservation technologies is frequently frustrated by institutional barriers. Conservation alone is not the reason that efficient technologies or processes are implemented. Many of the most important conservation advances are not related to new products and do not necessarily increase the market share of firms. In addition, information is imperfect, often unattainable in an unbiased form, and sometimes contrary to the interests of manufacturers. Moreover, in a number of situations the parties who make energy-related capital decisions are not those who must later pay the energy bills; for example, developers and occupants or landlords and tenants. To the extent that the appropriate market response to higher energy prices requires investment by firms or individuals, the efficient functioning of the market requires that these firms and individuals have access to the necessary capital. If they do not, market forces alone will not bring about an efficient response, and federal R&D can be appropriate to promote such efficiency.

Since the mid-1960s, there has been agreement that technological innovation is a key to growth in productivity and international competitiveness. Technical innovation during 1948-1966 accounted for 40 percent of productivity growth in the United States. Major governmental policies responded to this fact through the support of fundamental or basic research. Until the 1970s, however, the government supported applied research and development only where its own area of responsibility was involved or when it would be the user of the research results. While there have been commercial spinoffs from such government-sponsored R&D, little investment was made in the type of R&D that would specifically lead to new products and processes to be utilized by end-use sectors. The federal government's response to the energy crisis signaled a change in this direction. The Energy Research and Development Administration (ERDA) was created in 1974 to develop such products and energy technologies. For the first time an agency was to perform R&D through the commercialization stage on products and processes for which the government would not be the sole customer or utilize the results for enforcement of regulations. With this innovation came the requirement of establishing criteria for federal energy R&D in general and for conservation R&D in particular.

Priority for federal funding of R&D has been given to projects where the risks are so great (but potential rewards also great), the time for commercialization is so long, or potential returns based on present energy price are so low that private investment alone cannot rationally be expected to be adequate. Operationalizing this priority means that the following criteria should be used to select conservation R&D projects.

1. Risk. There are several types of risk. There are technical risks; e.g., will the new equipment or process work? There are economic risks; e.g., will heat pump water heaters sell if their first cost is twice that of a standard water heater? There are also risks associated with the acceptance of new technologies or systems where nontechnical or noneconomic barriers thwart their employment. Conservation products and services are especially subject to the latter forms of risks.

2. Payoff. Private investment is unlikely without a financial award or payoff. In the case of conservation technologies, the implementation of one unit may not have a high payoff, but there is the opportunity for many units to penetrate the market and thus save much energy. High payoff should also include the type of fuel to be saved by more efficiency or fuel switching. For example, oil is the commodity over which there is most concern, and priority should be given to those products which will reduce or eliminate oil consumption.

3. Time Factor. Will federally sponsored R&D accelerate entry into the market of new technology? By how many years? Will it accelerate market penetration? In order to accelerate entry into the market, a decision has to be made as to the scope of R&D projects. If federal dollars are used for research and development, are additional federal funds needed to demonstrate the technology on a pilot line as in an industrial process, or in an automobile engine? In some cases, there may need to be more than one "proof-of-concept" or

demonstration due to climate and geographical differences. Once the federal government's role in R&D is completed, the information needs to be transferred to manufacturers and users. Provisions and funds in any research program must be made for technology transfer. Where possible this should be done in concert with the user, such as jointly with a trade association (e.g., Iron and Steel Institute) or a professional association (e.g., Building Owners and Manager Association).

4. Cost Sharing. As research proceeds through the development chain, the private sector should assume more of the financial risk. If the private sector participates in the funding, there is more likelihood that the technology will be developed in a form that will be marketable.

5. Wide Applicability. Federally funded R&D should benefit more than one firm and have wide applicability. The former can be encouraged by supporting duplicate approaches up to a certain stage in the research, minimizing the amount of proprietary data (though this can have cost-sharing implications), and sponsoring appropriate workshops for the dissemination of results.

6. Support of Regulations. While it is important to distinguish R&D from regulation development, effective regulation frequently depends upon sound data for judging likely impacts of new technologies. For this reason, R&D is an essential component of regulation development. Yet, this effort can be controversial and difficult. Often there is a fear that industry will not participate in a federal agency's technological R&D if that same agency will subsequently regulate the industry.

7. National Impacts. Impacts on other areas of national concern - such as productivity, national security, competition, employment, materials, environment (both indoor and outdoor pollution) - must be evaluated.

These criteria need to be applied to each project, but they can never be applied simply through a formal quantitative index. In each case policy and technical judgment needs to be exercised once the criteria are used to identify reasonable candidates. A threshold criteria system was developed and used successfully for conservation programs at ERDA, and aspects of it are still in use today. The continuing limitation on this system, however, has been the lack of good data on such items as market potential and rate of saturation.

A comprehensive conservation R&D program would not only meet the criteria mentioned above but also that of balanced investment in various areas of conservation R&D. In addition to technological research, there would be investment in behavioral, institutional and health-related research. The overall program would be balanced in that some innovations would be ready for application and commercialization in the near term, and others in the middle and long term. In addition, technology transfer activities would be an important part of each R&D project, as would program evaluations to determine what has and has not been successful and why.

The Evolution of Federal Conservation R&D Programs

From 1973 to 1981, federal conservation R&D programs grew rapidly. Results from these programs are still being effective as the market penetration of new energy-efficient products continues to occur. With the Reagan administration, however, the philosophy toward federal sponsorship of conservation R&D changed dramatically.

During 1973-74, the federal government implemented various measures to minimize the impact of the energy embargo: lowering the speed limit to 55 mph, extending daylight saving time, closing gasoline stations on Sunday, lowering thermostats, removing lamps in office buildings. After the embargo, however, there was concern regarding U.S. strategies, particularly their vulnerability to the increasing impact of price escalation. In response, the federal government acted over several years to establish broad-ranging energy conservation programs in research and development, financial assistance and incentives, information, and regulation and deregulation of price, all of which were aimed at reducing energy consumption in the major end-use sectors - industry, transportation, and buildings.

Prior to the embargo of 1973, the federal government spent less than $3.0 million on conservation R&D. Most of this funding was from the National Science Foundation, Research Applied to National Needs program (NSF-RANN) and the National Bureau of Standards, Center for Building Technology. Research focused on the development of data, basic information, and models on how energy had been used historically and was then being used in buildings, industry and transportation. One of the earliest projects, the Princeton University - Twin River Project sponsored by NSF in 1972 and continuing today, initially monitored the construction of town houses and measured their energy use. Techniques such as infrared thermography were utilized to determine heat loss. A recognition of the importance of behavior as a part of conservation R&D resulted from these studies, which showed that the energy consumption of families with identical demographics (age, size of family, education, income) in identically built houses differed by a factor of 2.5 (Socolow, 1978).

Four months prior to the oil embargo, President Nixon announced that within the Department of Interior, an Office of Energy Conservation would be formed to coordinate an intended 7 percent reduction in federal energy consumption. However, there were no funds appropriated for this new office. As a result of the embargo, the Federal Energy Office and subsequently the Federal Energy Administration (FEA) were created. Incorporated into FEA, the Office of Conservation's first budget for FY 1974 was $1.5 million and was to fund all conservation activities, not just R&D. The House Appropriation report of June 1973 stated that "vigorous research in energy use, resource management and conservation is vital if government and industry are to provide the nation with a sustained and reliable energy supply in the future..." (House and Senate Committee Reports, 1975-84). A five-year conservation R&D plan was developed with assistance from the Departments of Housing and Urban Development and of Transportation, the National Bureau of Standards (NBS), and the General Services Administration. The activities funded at FEA were generally funded information programs and some mandatory state grants programs. Research funding also continued for the development of a data base to measure how energy is used in various sectors, to determine the costs and benefits of various retrofit options, to evaluate energy savings during the embargo, and to obtain information on residential conservation.

The creation of the Energy Research and Development Administration (ERDA) in 1974 signaled the intent to develop technologies for both the supply and the use of energy. These efforts were to be undertaken jointly with industry and were to be utilized by both federal and nonfederal sectors. Existing programs for energy supply (fossil, nuclear, oil and gas, geothermal, solar) were transferred to ERDA from other agencies. The Federal Nonnuclear Energy Research and Development Act of 1974 (PL 93-577) directed ERDA to initiate a national energy conservation research, development, and demonstration program. This was the first legislated mandate that there be federally supported R&D in building, industry and transportation energy efficiency. ERDA was faced with the task of establishing R&D programs responsive to the mandate, but it received little additional funding for new programs. Moreover, it incorporated a nuclear program of close to $1 billion which had previously been pursued for over thirty years at the Atomic Energy Commission. The only conservation program already in existence (the EPA vehicle heat-engine program which was developing alternative engines for pollution control) was transferred to ERDA (FY 1974 budget $3.0 million).

ERDA was required to submit budgets for FY 1975 and FY 1976 before it was fully operational (January 1975) and before a required national energy plan was prepared (June 1975) (see ERDA, 1975). For FY 1975, transportation received most of the conservation R&D funds (Table 1). By contrast in FY 1976, all of the end-use areas were funded, reflecting congressional sentiment that conservation was important. There was also sentiment that the growth and size of the program must be developed with prudent management and adequate planning. Except for transportation, there had been no management for the end-use area programs at ERDA until late April 1975. Congress was concerned that there would be adequate coordination between all the programs, particularly between ERDA and FEA (where funds were being used for mandated assistance programs to low-income households and to fund state energy conservation initiatives). The House Interior Subcommittee requested the investigative staff to conduct a study of energy conservation programs of FEA and ERDA with a view to identifying areas of overlap and duplication. During subsequent years the Office of Technology Assessment conducted several reviews of the conservation budget and continually recommended an aggressive research and development program for both technological and institutional innovation.

Table 1
ERDA Conservation R&D Funding FY 1975 - FY 1977[a]
($ in millions)

	FY 1975	FY 1976	FY[b] 1977
Buildings and Community Systems	2.4	11.5	35.6
Industry	-	5.2	16.0
Transportation	8.1	12.9	28.4
Multi-sector	2.3	7.3	24.4[c]
Total	12.8	36.9	104.4

Notes: [a]Throughout these charts, storage and electric energy systems, though organizationally part of conservation, are not included.

[b]The 3-month transition quarter is omitted, but it was .25 of FY 1976.

[c]Includes energy extension service which in future years is transferred to state and local programs.

ERDA submitted an energy R&D national plan to Congress in June 1975. The plan identified major changes that were needed in the nature and scope of the nation's energy R&D. These included an

> [i]mmediate focus on conservation efforts. These efforts implement first generation existing technology, extend this technology with improved capabilities, demonstrate its viability and widely disseminate the results. The primary targets are automotive, transportation, buildings and industrial process.

With support from the Ford administration and the Congress, the conservation R&D budget grew at a phenomenal rate, from $12.8 million in FY 1975 to $104.4 million in FY 1977. In the FY 1977 hearings, after ERDA had been operational for about a year, Congress approved more than a doubling of funding in each of the end-use areas. The House report in FY 1977 stated, however, that "ERDA has still to formulate a well-conceived long-range program. It is true that the program has been in operation for one year on an accelerated basis, but it is time progress was made." Both the House and Senate appropriations committees supported research, but not at a level as high as the authorization committees recommended.

The end-use programs developed at ERDA formed the basis for conservation efforts over the next nine years. The ERDA conservation programs were research oriented and were not involved with the implementation of regulatory programs or with mandatory deadlines. Behavioral and institutional research was initiated, particularly in the building sectors, as a supplement to technological research.

When President Carter entered the White House, ERDA and FEA were folded into what became the Department of Energy (DOE). The conservation programs from both organizations were merged into one office and organized by sectors. The ERDA program formed the basis for these programs with the added components from FEA of information, regulation and grant programs. The R&D programs continued to increase 40 percent annually, from FY 1978 to FY 1980, leveling off in FY 1981 (Table 2).

In addition to the research and development programs, there was increased emphasis on commercialization, field tests, and demonstrations. Programs in all stages of technology development (e.g., basic research, applied) were implemented. In addition nontechnological, behavioral and institutional research was conducted, including experimentation with delivery mechanisms. In creating the Department of

Energy in 1977, the Congress requested a sunset review of all the Department's programs after five years. A report was submitted to Congress to satisfy this requirement in February 1982, and that document summarizes the objectives and the accomplishments of the end-use program during this growth period (DOE, 1982).

Table 2
DOE Conservation R&D Funding FY 1978 - FY 1981
($ in millions)

	FY 1978	FY 1979	FY 1980	FY 1981
Buildings and Community Systems	63.4	79.3	103.4	91.3
Industry	31.9	40.0	60.2	69.6
Transportation	67.3	99.1	113.4	105.1
Multi-sector	3.0	10.2	19.1	34.1
Total	165.6	228.6	296.1	300.1

Unlike the Ford and Carter administrations, which were implementing a balanced R&D program, the Reagan administration proposed budgets with huge reductions for all technology R&D except nuclear. In February 1981, one month after taking office (five months into the fiscal year), Reagan submitted a request to rescind money for conservation appropriated for FY 1981 and submitted his own 1982 request (Table 3). From FY 1982 through FY 1984 the Reagan administration proposed elimination of all conservation R&D programs except for a few long-term, high-risk research programs (Table 4). The Reagan philosophy was that the private sector would conduct virtually all the research needed and that the federal government should not supplant the marketplace in areas of development and demonstration. These views were not supported by impartial analysis intended to evaluate the costs and benefits of the conservation program (Energy Research Advisory Board, 1983). Rather, the attempt to eliminate conservation and renewable programs was guided almost entirely by ideology and by assumptions that the private sector would do all the necessary research and development and implementation of conservation programs. Consistent with this view, the Reagan administration also attempted to eliminate mandatory programs through the budget process, rather than request repeal of the enabling legislation. In the FY 1983 budget request, the administration stated that if no funds were appropriated for programs such as appliance efficiency standards, residential conservation services, and industrial reporting, the legislated mandates would not need to be followed.

Table 3
Reagan Rescission and Revised Budget Request
($ in millions)

	Carter		Reagan		Final Appropriation	
	FY 1981	FY 1982	FY 1981	FY 1982	FY 1981	FY 1982
Buildings and Community Systems	113.8[a]	130.6	62.0	31.1	64.2	42.8
Industry	89.7[a]	51.6	43.0	0.9	41.2	28.8
Transportation	119.7[a]	121.7	81.0	38.3	92.9	58.9
Multi-sector	8.0[b]	12.1	8.0	12.1	8.2	7.8
Total	331.2	316.0	194.0	82.4	206.5	138.3

Notes: [a]From U.S. Senate.
[b]Energy Conversion and Utilization Technologies Program only.

Table 4
Reagan Administration Request for DOE Conservation R&D
FY 1982 - 1985
($ in millions)

	FY 1982	FY 1983	FY 1984	FY 1985
Building and Community System	31.1	-	16.4	35.3
Industry	0.9	-	10.6	33.3
Transportation	38.3	-	31.6	51.1
Multi-sector	12.1	17.0[a]	3.0	21.5
Total	82.4	17.0	61.6	141.2

Notes: [a]Long-term, generic research across all sectors.

Both the Republican Senate and Democratic House overwhelmingly rejected Reagan's conservation R&D budgets and appropriated larger amounts (Table 5) with very specific programmatic guidance in the conference reports and mandated personnel ceilings to manage the programs. In FY 1983, Reagan proposed only $17 million for conservation R&D and proposed to increase it to $61.6 million in FY 1984. The Secretary of Energy boasted that this increase for FY 1984 was four times what was requested the previous year, but Senator McClure (R. Idaho) pointed out that this request represented a 78.4 percent reduction from the fiscal 1983 appropriated program level. No other area of DOE's budget had sustained such a reduction. For FY 1985, Reagan has requested a conservation budget of $141 million, which is more comparable to the congressional appropriation of FY 1984. In this budget, nuclear fission and fusion (the only DOE programs with requests approximately at the FY 1981 levels) dominate the budget, comprising two-thirds of the energy technologies budget request compared to just one-third in 1981. Energy conservation accounts for only 17 percent, fossil energy accounts for only 12 percent, and solar energy accounts for only 8 percent of the energy technologies portion of the budget. In the FY 1985 budget proposal, nuclear defense activities would receive 62 percent of the total DOE budget (excluding the Strategic Petroleum Reserve). This is a $4.2 billion increase over the FY 1981 nuclear weapons budget and more than $1 billion over FY 1984. Reagan proposed to reduce the civilian energy share to 32 percent of the DOE budget (DOE, 1984a). In addition, conservation has been able to prepare multiyear planning documents (DOE, 1984d).

Table 5
Congressional Appropriation for DOE Conservation R&D
FY 1981 - FY 1984
($ in millions)

	FY 1981(R)	FY 1982	FY 1983	FY 1984
Building and Community Systems	64.2	42.8	33.5	38.4
Industry	41.2	28.8	27.9	33.7
Transportation	92.9	58.9	54.8	65.0
Multi-sector	8.2	7.8	9.2	9.3
Total	206.5	138.3	125.4	142.4

A Sectoral Analysis of Federal Conservation R&D Programs

Despite the recent challenges to conservation R&D funding, federal programs in this area have been characterized by substantial accomplishment. The scope of these accomplishments can be recognized by a review of programs by sectors. The U.S. consumed 73.7 quads of energy in 1984: 35 percent was consumed in the residential/commercial sector; 38 percent in the industrial sector; and 27 percent in the transportation sector. The first three sections briefly describe the characteristics of each sector, provide examples of successful R&D programs, and suggest future needs for R&D investment consistent with projected energy use. A final section looks at energy conversion and utilization technologies.

Buildings Sector

Residential and commercial buildings used approximately 26.0 quads of primary energy in the U.S. in 1984. While energy demand by this sector has changed little over the past ten years (24.5 quads in 1973), the percentage of national energy consumed by buildings has risen from 33 percent in 1973 to its present 35 percent. A recent study by the Office of Technology Assessment of buildings in cities concluded that only 38 percent of the energy savings achievable by investments that are economically justified under present conditions of capital, retrofit information and public policy will in fact occur by the year 2000 (OTA, 1982). The building industry is extremely fragmented and hard to influence through any single coordinated series of activities. It is also less responsive than the other sectors to rising prices. In this industry, the risk and cost of nonperformance or subperformance can be large to the architect, engineer, banker, and developer. This predicates a design practice based on what works, what the designer has direct experience with either in his own practice or from peer sources. Consequently, the rate of diffusion of new designs, construction practices, and technologies is slow. Studies have shown that the level of privately sponsored research and development in buildings is less than in the other end-use sectors.

Once built, homes and offices are turned over to families, maintenance people, and tenants whose decisions and actions have an enormous impact on energy consumption. Yet those who are asked to operate and maintain the building stock typically have little information or training on energy conservation. Surveys consistently reveal, for example, that Americans believe the most important energy conservation action one can take is to turn off the lights or the television, yet these make relatively small contributions to energy savings (Kempton, 1982; Stern, 1983). Although the utilities through the Residential Conservation Program are providing diagnostic services to homeowners on how to make cost-effective retrofits, there is no well-developed industry for putting energy conservation improvements in place in existing homes (the so-called "retrofit" market). Finally, approximately one-third of the housing stock is rented rather than occupied by the owner. In such housing, the incentives to conserve are split between landlord and tenant in a fashion that raises substantial institutional problems. It is appropriate that R&D investments be made to address these problems.

Successful conservation R&D programs for buildings have led to new products being introduced into the marketplace and new methods to audit and retrofit existing buildings. For example, the DOE Consumer Products Program has funded the cost-shared development of the first successful heat-pump water heater and thereby created a new industry. The heat-pump water heater is an appliance that uses the refrigeration cycle to heat water and requires only half the electricity of conventional resistance heating. Two small companies were funded by DOE for market development, field demonstration, and testing of an integrated heat-pump water heater (DOE contributed $334,000 and EUS and E-Tech provided $200,000). The full-scale demonstration included several utilities' agreeing to purchase and install at least five models and collect data in a uniform manner determined by scientists at Oak Ridge National Laboratory (ORNL). As a result of such tests and the data obtained and published by ORNL, the two small companies decided to manufacture the heat-pump water heater. By the end of 1982, there were eleven manufacturers of the heat-pump water heaters listed in the GAMA directory. The larger manufacturers who initiated production stated that they did so as a result of the DOE test data. Consumers, however, are not purchasing the product as quickly as anticipated due to their higher initial cost relative to conventional water heaters.

Other successful federally sponsored demonstrations have led to the introduction of a high-efficiency refrigerator-freezer that is 40 percent more efficient than earlier models, to improved solid-state lighting ballasts, to advanced oil burners, and to an advanced motor-compressor for refrigeration. In addition to conducting technology R&D, there have been programs to remove institutional barriers and to provide innovative financing and information. For example, a demonstration project in Lakewood, New Jersey, was developed for an energy services company to provide no-cost audits to the consumer and retrofits on one thousand houses. The consumers participating in the demonstration included a retirement community (all electric) and a township representative of the area (blue-collar families with median income of $21,000 in single, detached houses). The energy services company contracted with the utilities to sell them "saved energy" resulting from the retrofits. The contract stipulated that only energy actually saved would be paid for by the utility. Several important findings emerged. There was considerable interest in the program exhibited by the communities. The retirement community, for example, heard about the program and actively sought to participate. Several meetings were held to explain the program with little money spent on an extensive household-to-household marketing campaign. In fact, within several days of the follow-up meetings, nearly every household in the retirement community had signed up for the service, and the energy services company had to expand the number served. Although the savings were not as high as projected, there was a 10 percent reduction in utility bills. Most notably, households were satisfied and noted an improvement in their comfort level. While there were start-up lessons learned from this demonstration in Lakewood on the need for quality control in retrofit installations, the important finding was that the concept could work (Mellon, 1983). The company that participated in the demonstration (putting in one-third of the funds) and two other companies and the utility have signed contracts for additional projects.

Shared savings or performance contracting approaches, in which building owners are not required to capitalize the energy-efficiency improvements since a contractor or third party capitalizes the investments, have become a new method to finance retrofit of buildings, particularly institutional buildings (schools, hospitals, not-for-profits, government buildings). There are now at least 75 energy service companies providing this arrangement, whereas in 1980 there were fewer than 10. Several of the companies offering shared savings, particularly for smaller buildings (less than $50,000 in yearly energy costs) state that they were able to develop business-as-usual agreements and lease arrangements based on model agreements developed by the DOE program. The development and testing of model "paper" tools and other innovative institutional and organizational arrangements is as important as testing out actual hardware.

In the community systems area, DOE in the past has provided funds for feasibility studies and preliminary design for integrated community energy systems and district heating. Several of these have gone forward to construction without major federal dollars (Trenton, New Jersey; St. Paul, Minnesota). Without the front-end DOE support, however, they would not have proceeded. These grants are relatively inexpensive (less than 1.0 million for district heating feasibility studies in 28 cities), and they accelerate the introduction of energy-efficient delivery systems at the community level.

Space heating and cooling are the major users of energy in the building sector, with water heating important in the residential sector and lighting in the commercial sector. Energy use projections for the year 2000 indicate energy for space cooling will increase the most, with some increases in water heating and decreases in space heating and lighting. Thus, research and demonstration in the next decade should continue to focus on space conditioning equipment, water heating, and refrigerants in addition to institutional and information questions. More work is also needed on multi-family and commercial building energy-efficient technologies and their implementation; these types of buildings have received less attention than single-family residential buildings.

It is important to obtain better information and data as to how new energy-efficient residential and commercial building systems operate both technically and economically. How do these innovations meet design specifications? How do various subsystems interact? How do the occupants respond to these new buildings? Data then needs to be analyzed and results conveyed to architects, builders, and developers in formats they understand to accelerate widespread adoption of the most successful and energy-efficient designs and practices. Some residential data is collected through the Energy Information Administration Consumption Surveys and the LBL-BECA system, and an effort has recently been initiated with the American Institute of Architects Foundation to do post-occupancy evaluation of lighting in office buildings, but there is no comprehensive data collection system (EIA, 1983a, b, c, d; 1978).

Good engineering data not only should be obtained for new buildings, but also for retrofits of multi-family and commercial buildings. In addition, methods to accomplish retrofits, particularly in multi-family buildings, need to be demonstrated. For the next twenty years, retrofits will continue to be important, but there has been little documentation of the costs and benefits of specific measures or of the interactive effectiveness of different measures except in single-family housing. In the building area, the federal government should use its own facilities to demonstrate new technologies and new approaches and to support production through government purchases.

Industrial Sector

In 1984, the industrial sector consumed 27.9 quads of energy. The basic materials industries - chemicals, primary metals, petroleum refining, paper making, clay and glass - consume about 78 percent of all the energy in this sector. In these industries, energy costs average about 30 percent of the value added; the industry-wide average is 10 percent. The industrial share of U.S. energy has declined from 42 percent in 1972 to 38 percent in 1984. Industrial energy consumption in this period may have been more price responsive than in the rest of the economy. Energy consumption per unit of production fell by almost 21 percent during the period 1972-1982. This reduction was accomplished through capital replacement, accelerated implementation of energy-efficient practices and equipment, and shifts towards less energy-intensive product mixes. Production growth in manufacturing and other industries kept pace with the efficiency gain during this period. In 1982 and 1983, the annual energy consumption was 4.0 quads less than in 1980, which is attributed to the economic recession and could increase with economic recovery. In fact, in 1984 the industrial energy consumption increased by 2.0 quads from that in 1983 (EIA, 1985).

The DOE energy conservation program for industry is organized into four areas: waste energy reduction (which applies to many industries), process efficiency (which applies to specific industry), advanced cogeneration technologies, and technology transfer activities. Since its initiation in 1975, the industrial program has required extensive justification for federal funding because of the belief by both Republicans and Democrats that "industry is better able to do it itself." In fact, industry's focus has been on short-term returns on investment. It has not funded longer-term, high-risk energy-efficient technology developments, particularly where no increase in market share or new product was involved. The Carter administration and Congress funded the industrial program to a credible size, and since the inception of the program, industry has approached the government with worthwhile cost-shared projects. During the first three Reagan years, however, the administration proposed to eliminate the program. During this same period, unsolicited proposals with new ideas from industry diminished, and industry did not begin to fund projects on its own (Hagler, 1984).

The industrial conservation program (IP) has demonstrated some of the most successful projects in terms of technology and implementation. There has been continued tracking of the implementation of technologies and processes that have been funded, and studies have been conducted to analyze factors influencing the success of projects. There has also been utilization of innovative contracting procedures to recover the costs of demonstration. Over 23 projects have been completed by the DOE industrial program and are currently saving *each year* 21.4 trillion Btu's ($85.6 million at $4 per million Btu) at a total federal cost of $34.3 million. An analysis performed in 1981 to identify factors common to projects considered to be successful found the following characteristics (Hagler, 1981):

o The most significant factor was the contractor's incentive for conducting the project. Projects that were undertaken with a market/sales objective generally proceeded more rapidly to market introduction with fewer cost overruns. Specifically, many of the more successful projects were undertaken by contractors seeking to develop a product that they intended to sell.

o The type of contractor that tended to be most capable of bringing a technology to the market introduction stage was a large, technically competent, well-respected, and innovative equipment manufacturer. Moreover, projects selected for funding partially based on the contractor's good reputation tended to be more successful than projects for which the contractor's reputation was not considered in the proposal evaluation.

This analysis, and others such as the National Academy of Science, have underscored the necessity for good technology transfer and understanding of market potential (National Materials Advisory Board, 1981, 1982, 1983). Successful projects involve a technology that will fill an expressed market need, that is developed by a contractor who plans to sell the resulting product, and that is cost-shared by the contractor with the federal government. Where a specific industry is not developing the technology, interaction with trade associations at an early point can help insure effective transfer.

There are cases of successful federal-industrial demonstration projects. For example, a cost-shared program between DOE, Thermo-Electron and Hoffman-LaRoche was initiated in October 1979 for the first slow-speed diesel cogeneration system in the United States. The objective was to demonstrate the technical and economic viability of the system to simultaneously provide electric power and process steam in a pharmaceuticals plant. The system began operations in November 1982, producing 23.3 MW of electricity, 160,000 pounds per hour of steam, and 262,000 pounds per hour of 170 degrees Fahrenheit boiler feedwater. This single installation is saving 1.5 trillion Btu's annually, and four similar installations are being negotiated without federal involvement. The total cost was $22,400,000; DOE's cost share was $8,745,000.

In 1979-80, several projects were intended to be tested at various host sites with the objective of accelerating the implementation of the technologies. Implementation of multiple demonstration sites did not go as rapidly as intended due to procurement problems. These products are now penetrating the market through the private sector, and it is questionable whether multiple testing was even necessary to give them a boost. Tests in various environments widen markets, but in most of the IP projects a single, representative, full-scale demonstration appears to be adequate for technology transfer.

An important part of the DOE industrial program is rapid dissemination of results to potential users. This is accomplished by cost-sharing projects with firms interested in implementing the technology, workshops to brief potential users and manufacturers, and dissemination through appropriate trade and professional associations. Small- and medium-size manufacturers often need energy management assistance. The Energy Analysis and Diagnostic Center (EADC) program utilizes students at local universities supervised by engineering faculty to perform energy audits for these groups. To date, over 700 manufacturers in seven states have been assisted with 50 percent of the recommendations being implemented.

Projected fuel and electricity use in the year 2000 indicates that certain industry groups are expected to grow more rapidly than the industrial sector as a whole. These include chemicals; stone, clay and glass; machinery; other manufacturing industries; and nonmanufacturing industries. Primary metals, textiles, petroleum, and fabricated metals will probably be slowest. Part of this expectation is due to the increasing dependence on foreign sources for strategic materials or manufactured equipment and final products. Much of the improved processing used in traditional industry has come from foreign sources. In future years, there will be opportunities to capitalize on the developing technology base associated with sensors, microprocessors, and computer control and their application. The U.S. must decide whether to retain the "smokestack" industries and manufacture our own textiles, steel and aluminum, or to purchase "off-shore" and become more of an information economy. Should the objective be to have American industry in the forefront of manufacturing in traditional and new industries, the cost-shared development and demonstration of energy-efficient processes and waste energy reduction techniques must be continued. The recent expansion of the thin strip steel casting project to the demonstration phase is an example (Savitz, 1984). This project will increase the productivity of producing metal strip, and if energy efficiency improvements are proven through laboratory-scale experiments, a joint industry-government, pilot-scale test should be undertaken. There are few examples of industrial conservation research programs initiated during the past three years. Much of the current program consists of earlier starts that are progressing to the follow-on stage after successful completion of an initial phase.

Transportation Sector

The transportation sector consumed 27 percent of the nation's energy in 1984, but 61 percent of the nation's petroleum. Of the energy used in this sector, 97 percent is in the form of petroleum products. Use of petroleum in transportation has risen 10 percentage points during the past decade. The transportation sector is much different than the other two in that it is a concentrated industry and there is relatively rapid turnover of products. Moreover, it has been most dramatically impacted by imports. As a result of the Corporate Average Fuel Economy (CAFE) standards, the efficiency of automobiles has greatly improved in the past decade. The DOE conservation R&D program has focused on longer-term options. The two major programs, advanced heat-engine development and electric and hybrid vehicle development, are being developed as alternatives to the internal combustion engine.

In 1975, the Jet Propulsion Laboratory completed an analysis of alternative automotive power plants for possible introduction during the 1980-1990 time period (Jet Propulsion Laboratory, 1975). Alternatives to the internal combustion engine were evaluated from a societal point of view in terms of energy efficiency, air quality, consumer cost, materials availability, safety, and industry impact. The analysis showed that the goals for emission reduction and energy conservation for the automobile over the next five to ten years could be met by improvements to existing engines and vehicles. This would provide time for the necessary development work on the Brayton (gas turbine) and Stirling engines. The study showed these systems would require intensive component, system and manufacturing process development at a funding level ($150 million/year) considerably higher than the 1975 level for transportation programs ($5-13 million).

Several of the alternate power plants considered in the JPL study, if mass produced, would result in increased consumption of stainless steel and introduce significant use of super alloys into the automotive industry. No heat engine type was discarded due to materials availability, but configurations were selected with material cost and availability in mind. Ceramic materials were identified in the JPL study as a technology that could be used in lieu of metal alloys in the construction of actual components, and their introduction would ameliorate the demand for higher priced metals (nickel, chromium, cobalt, and tungsten) from foreign sources. Thus, research and development into ceramics material (components, manufacturing, characterization and processes) was necessary to obtain successful introduction on a wide scale. In 1975, ceramics were materials that went into sinks, dishes and bricks. By the late 1970s, widespread use of ceramics was occurring in the electronics industry, but that is not the same as making turbine wheels and rotors that can spin at 125,000 rpm at 1200 degrees Centigrade.

The JPL report and subsequent legislation (1978) laid the foundation for a joint industry-government preproduction proof-of-concept engine development program. These engines were to be mid-size (3,000 pounds), 30 percent better than the 1985 CAFE standards, meet all environmental requirements, and have multi-fuel capability. Industry would then provide the capital to retool for production and design. The program has focused on the development of advanced automotive gas turbine and Stirling engines for light-duty vehicles and uncooled, minimum friction diesel engines with integrated waste heat recovery for the heavy-duty transport sector.

Parallel to the engine development programs has been a research program on key technologies with specific emphasis on ceramics. In 1979, two gas turbine development programs were initiated - each to cost $60 million and each cost shared. Each of the programs involved the design, development, and testing of advanced automotive gas turbines incorporating ceramic components. Ceramic development has focused on proof-of-concept ceramic parts operating under automobile application conditions. Although a commercialized ceramic gas turbine engine is probably a decade away, ceramic turbochargers are being evaluated both here and in Japan for introduction within the next two years. In the U.S. the development of a ceramic turbocharger is being done without federal funds, but the project is utilizing knowledge gained in the federal program. While the development project has been very successful, additional fundamental research activities were required before an industrial technology base was established which would allow future production of reliable ceramic components commercially. Consequently, a parallel basic research program was initiated which is investigating the ceramic materials and processing, design methodology, and data and life prediction. Thus, there has been both a long-term commitment to proof-of-concept automotive technology development and a continuing contribution to basic research. Both programs have reached the milestones scheduled with funding available. If there had been a focus only on basic research with no identified product or only on the proof-of-concept product, it is unlikely the U.S. would currently be competitive today with Japan in ceramics technology.

The Stirling engine development project has also made progress. A USA-built Mod 1 automotive Stirling engine has been characterized on an engine dynamometer, and engines are currently being evaluated by some manufacturers. There are several major R&D needs for the Stirling engine, including long-life seals and metal conditioning for hydrogen containment, advanced materials for light-weight and low-cost, and rapid-response control systems. A question that remains to be addressed is whether Stirling engine technology would initially be better suited for stationary power. There has also been research to develop a Stirling driven heat pump, but a recent assessment for best application of Stirling technology has not been performed.

The other major transportation project has been the Electric and Hybrid Program. The program involves research on two types of electric vehicles: 1) an all-electric commercial van with limited range, and 2) an all-electric passenger car with unlimited range. It has been demonstrated that an electric vehicle can be designed that can fully compete with contemporary gasoline and diesel fuel engines. The first of these is the General Electric/Chrysler Electric Vehicle (ETV-1). The vehicle seats four adults and weighs about 3000 pounds, including 1000 pounds of batteries (1982 batteries). The vehicle has a top speed in excess of 70 mph and an acceleration time of 8.6 seconds for 0.30 mph. The range of the vehicle is 124 miles at 35 mph and 74 miles for an urban cycle.

The major technology need in this area is a successful battery system. Research has focused on aqueous, nonflow batteries such as lead and nickel-iron, flow batteries such as zinc bromide, and high temperature batteries such as sodium sulfur. The extent of progress made in battery modules is given in Table 6. Other studies indicate that advanced propulsion systems can reduce the required battery energy density up to 40 percent compared to propulsion systems available in 1976. Light weight, integrated, advanced alternating current propulsion systems are also being evaluated in test bed vehicles.

Until 1981, there was a major demonstration program for electric and hybrid vehicles. The Electric and Hybrid Vehicle Research, Development and Demonstration Act of 1976 (PL 94-413) required the government to subsidize the deployment of 7,500 electric vehicles, 2,500 in FY 1978 and 5,000 spread over an additional four-year period. The act was subsequently modified to stretch out the purchase of the initial 2,500 vehicles over a three-year period. There was a parallel requirement in the act for the development of improved technology in battery and propulsion systems. From 1979 to 1981 about

1,100 vehicles were purchased, mainly from four small business manufacturers, with loan guarantees for two companies. These companies replaced the existing internal combustion engine of vehicles purchased from manufacturers with electric drive systems and batteries.

Table 6
Performance Objectives and Accomplishments of EV Aqueous Nonflow Battery Modules

Objective	Pre-Project Lead-Acid Status (1977)	FY 1983 Lead-Acid Accomplishments[a]	FY 1983 Nickel/Iron Accomplishments[a]	Stated Industry Goals
Specific Energy (Wh/kg)[b]	30	41	48[c]	56
Specific Peak Power (W/kg)[d]	96	104	103	104
Cycle Life (cycles)[e]	250	508	>904	800

Source: U.S. Department of Energy. 1983. *7th Annual Report to Congress for Fiscal Year 1983: Electric and Hybrid Vehicles Program.*

Notes: [a]Based on verified results demonstrated at the National Battery Test Laboratory for multi-cell modules.

[b]At the 3-h discharge rate.

[c]Preliminary tests on nickel/iron modules delivered in late FY 1983 demonstrated 57 Wh/kg; life tests are in process.

[d]Peak power (30-s average) at 50 percent state-of-charge.

[e]Cycling at 80 percent depth of discharge; life to 75 percent retained capacity.

It is in the national interest to develop electric vehicle technology since electricity can be generated from more abundant fuel sources and EVs can be a real alternative for petroleum-based transportation. However, the large-scale demonstration of 2,500 new technology vehicles as provided for in the original legislation was recognized by Congress as being premature. After an oversight hearing, the legislation was amended to stretch out the procurement over a three-year period to permit new technology to be incorporated into the later years' procurements. In FY 1981 it was decided to phase out the large-scale demonstration of existing technology vehicles and concentrate on developing battery and propulsion system technology and on testing and evaluating new technology in vehicle fleet operation. In hindsight, the large vehicle demonstration project was premature; technology development should have preceded any large-scale demonstration. Prototype vehicles should have been designed, built and tested with joint participation by industry.

Since the automobile has been the major user of transportation energy, it has been the focus of transportation energy conservation programs. Projections indicate that by the year 2000 energy use by trucks will be as large as that for automobiles. The R&D program over the next decade should look at technology development opportunities to make light and heavy trucks more efficient and capable of burning many fuels. Trucks cannot be downsized in the same manner as automobiles; therefore, new technology will be essential for conservation. Some heat engine technology is applicable to trucks, and some experimental programs have been initiated using ceramics and other technology. These projects need to be evaluated more systematically, and long-term joint development programs considered and implemented. Although the advanced gas turbine program is to be completed in 1986 and all ceramic engines will have been tested, there will be need for additional iterative research involving design for long-life engines and fabrication and test of ceramic components in an engine environment.

Energy Conversion and Utilization Technologies

In addition to research in specific end-use areas, it is important to expand the technology base and fundamental knowledge applicable to more than one sector. In 1981, the Energy Conversion and Utilization Technology program was initiated to develop technology for improved energy systems and devices. The program does not develop hardware, but conducts R&D that addresses specific technology deficiencies in the areas of combustion, materials, tribology, and catalysts/biocatalysis (DOE, 1981-82). This initiative has led to significant accomplishments in its short existence. One has been the application of new materials developed at national laboratories. Another has been in the combustion area, where an advanced combustion laboratory was built at Sandia National Laboratory with DOE funds. Corporate-sponsored scientists and academic researchers utilize the facilities' advanced laser measurement technique and large-scale, multidimensional computer models to better understand combustion engines and thereby increase fuel efficiency, control emissions, and make possible broader fuel use. The primary source of unburned hydrocarbons is raw fuel and oil trapped in crevice volumes and carbon deposits, rather than deposits in the cooler quench layers in the chamber as was previously believed. Now that this is well understood, auto industry research teams are developing ways to minimize this problem, which could result in the elimination of automotive catalytic converters.

The Challenge to the Conservation R&D Program

Review of DOE programs by ORNL and DOE's policy office have shown that federal conservation programs have been responsible for significant energy savings in absolute terms. Improvements in energy efficiency are estimated to account for about 14 quads of reduced growth in energy demand and about 60 quads of cumulative energy savings over the past ten years. Most of this decrease has taken place in the private sector. Probably the most influential factors in this shift have been energy prices and the threat of future shortages. Estimates by Oak Ridge National Laboratory and DOE suggest that government conservation programs resulted in as much as 20 percent to as little as 5 percent of the savings (Adams et al., 1984; Greene et al., 1982; Hirst et al., 1983; Hirst et al., 1981; DOE, 1984b; DOE, 1982). Thus, the savings to the U.S. economy are at a minimum $3.5 billion, and possibly as much as $14.0 billion in 1983 - up to five times as much as the government spent on energy conservation programs in the past ten years. These savings continue to accumulate and new improvements in energy efficiency are added annually.

The conservation R&D program has been one of the most successful, cost-effective energy programs that the government has sponsored. It has led to the development of high-risk technologies jointly funded with industry. Moreover, upon demonstration of technologies, the private sector has proceeded to implementation of energy efficient options with no additional federal funds required. Without the "seed money" from government, most industries would not have undertaken the risks of innovation. Reductions and elimination of government funds also eliminate, in some cases, the funds industry would spend in sharing the risk with government. In this context recent reductions in federal conservation funding pose the most serious threat to continued accomplishment.

The impact of reductions in program support is equally apparent in regard to retaining the infrastructure and personnel needed for program success. As energy conservation technologies became a legitimate R&D area in the 1970s, physicists, chemists, architects, engineers, and economists started to attack the problems; thus highly competent trained people began to apply their knowledge and techniques to solving these new problems. Unfortunately, the uncertainty of the programs the last few years has caused some of these people to do research in other areas. In addition, in September 1981, the Reagan administration ran a reduction in force (RIF) of headquarters people. Over one-third of the people in conservation were eliminated or left. Many of the best middle-level managers were transferred from areas of their expertise, looked for jobs outside the government, and left government. Conservation programs had been well managed, with no cost overruns, no schedule slippage except where funds stretched out programs, and significant accomplishments. As a result of the premature RIF (carried out because programs that the administration wanted eliminated were not canceled by Congress), Congress has mandated personnel floors which are still in effect. At the time of the FY 1986 budget submission, these floors still were not met by the administration.

Despite the challenges to the conservation program during the Reagan administration, the federal role in conservation R&D that evolved in the 1970s is still appropriate today. The most recent National Energy Plan acknowledges the importance of conservation as an option (DOE, 1983a). The federal role should continue to be developed to complement the marketplace and intervene where the market does not function or where there is market failure. Moreover, the objectives which guided the program in the past are still pertinent: (1) to accelerate the development of long-run economic efficiency in both the supply and use of energy; (2) to reduce key uncertainties about the nation's energy future by creating new energy options and insuring a smooth and equitable transition to domestically abundant fuel and renewable domestic energy sources; (3) to reduce the nation's vulnerability to oil supply disruption; and (4) to improve the competitive position of the U.S. in the world economy. The fourth objective has become important more recently as imports have increased in all the sectors of the economy. For example, ten years ago 10 percent of the automobiles were imported; today, 27 percent of the automobiles are imported and many U.S. automobiles have imported parts. This fourth objective is not unique for energy-efficient technologies.

Ten years after these programs have begun, it is appropriate to review the technology needs of the 1990s. One focus of this review should be the efficiency of current products and production processes and the basic research necessary to create conditions for future efficiency. Research and development activities should continue to support incremental changes, but there should also be support for projects that could lead to revolutionary advances and major structural changes. There need to be continual and long-term commitments as long as there is legitimate research progress. If policies are inconsistent and vary from year to year - as in the past four years - industry may be reluctant to enter into relationships with government. Alternatives to annual appropriations with provisions for oversight need to be explored.

Cost-shared research between government and industry is important. Investment by others besides the federal government can help insure more rapid commercialization. There does not need to be a fixed formula, but as the technology moves from one development phase to the next and the risks become reduced, the private sector share should increase. The appropriate level of cost sharing would be determined on a case-by-case basis and be influenced by such factors as the ability of the proposer to share; the vested interest of the proposer; the residual rights required by the proposer; and the level of risk of the project. Innovative payback schemes - such as refunding savings from government-supported projects (e.g., coil coating) or royalty payments to the government from a certain number of sales - could be created to recover the federal share of project costs.

We live in a world of intense competition. Progress needs to be continued toward an energy-efficient economy in order for the U.S. to remain competitive. In this regard, it is important for the U.S. to diversify its energy base so that it has access to secure and alternative energy sources at affordable prices; and it must develop technologies which enhance long-term efficiency. The federal-private partnership in research and demonstration of end-use technologies and implementation strategies must be maintained. If adequate funding is appropriated, conservation R&D initiatives will continue to contribute to an energy-efficient future. Conservation R&D programs need to be evaluated and pursued in an environment that extends beyond ideology.

REFERENCES

Adams, R. C. et al. 1984. *A Retrospective Analysis of Energy Use and Conservation Trends. 1972-1982.* Pacific Northwest Laboratory, Department of Energy Contract DE-AC06-76RLO 1830 (August).

DOE (U.S. Department of Energy). 1980-85. Budget Submissions, FY 1980-85.

________________. 1984a. FY 1985 Budget Highlights (February).

______________. 1984b. Office of Policy, Safety and Environment. "Energy Use Trends in the United States 1972-82," n.p. (February 28).

______________. 1984c. Office of Industrial Programs. *Annual Report to Congress and to President on 1982 Industrial Energy Efficiency Improvement* (April). Washington, DC. (Available from the Office of Industrial Programs.)

______________. 1984d. Office of Conservation. *Energy Conservation Multi-Year FY 1986-FY 1990* (July).

______________. 1983a. *The National Energy Policy Plan* (October). Washington, DC: U.S. Government Printing Office.

______________. 1983b. Office of Policy Planning and Analysis. *Energy Projections to the Year 2010: A Technical Report in Support of the National Energy Plan* (October). Washington, DC.

______________. 1983c. *7th Annual Report to Congress for Fiscal Year 1983: Electric and Hybrid Vehicles Program.* Washington, DC: U.S. Government Printing Office.

______________. 1982. *Report to Congress: Sunset Review* (February).

______________. 1981-82. *Annual Report, 1981-1982.* Washington, DC: U.S. Government Printing Office.

EIA (Energy Information Administration). 1985. U.S. Department of Energy. *Monthly Energy Review.* DOE/EIA-0035 (84/12[4]) (March). Washington, DC: U.S. Government Printing Office.

______________. 1984. U.S. Department of Energy. *Annual Report of Energy Conservation Indicators for FY 1982.* DOE/EIA-0441(82) (January). Washington, DC: U.S. Government Printing Office.

______________. 1983a. U.S. Department of Energy. *Residential Energy Consumption Survey: Consumption and Expenditures - April 1980 Through March 1981.* DOE/EIA-0321/2 (June). Washington, DC: U.S. Government Printing Office.

______________. 1983b. U.S. Department of Energy. *Energy Price and Expenditure Data Report, 1970-1980 State and U.S. Total.* DOE/EIA-0376 (July). Washington, DC: U.S. Government Printing Office.

______________. 1983c. U.S. Department of Energy. *Residential Energy Consumption Survey: Housing Characteristics 1981.* DOE/EIA-0314(81) (August). Washington, DC: U.S. Government Printing Office.

______________. 1983d. U.S. Department of Energy.

Residential Energy Consumption Survey: Consumption and Expenditures - April 1981 through March 1982. DOE/EIA-0321/1(81) (September). Washington, DC: U.S. Government Printing Office.

_______________. 1978. U.S. Department of Energy. *Interim Survey of Energy Consumption in U.S. Households.* Washington, DC: U.S. Government Printing Office.

Energy Research Advisory Board. 1983. *Energy Conservation and the Federal Government: Research, Development, and Management* (January). Washington, DC: U.S. Government Printing Office.

ERDA (Energy Research and Development Administration). 1975. *A National Plan for Energy Research, Development and Demonstration: Creating Energy Choices for the Future,* ERDA-48 (June).

Greene, D., E. Hirst, J. Soderstrom, and J. Trimble. 1982. *Estimating the Total Impact on Energy Consumption of Department of Energy Construction Program.* ORNL-5925 (November).

Hagler, Bailly and Company. 1984. *Industry Participation in R&D.* Report to the Director, Office of Industrial Programs, U.S. Department of Energy (November).

Hagler, Bailly and Company. 1981. *Analysis of Factors Influencing Success of OIP Projects.* Memorandum to the Director, Office of Industrial Programs, U.S. Department of Energy (December).

Hirst, Eric, R. Marlay, D. L. Greene, and R. W. Barnes. 1983. "Recent Changes in U.S. Energy Consumption: What Happened and Why." *Annual Review of Energy* 8 (February): 193-245.

Hirst, Eric, R. W. Barnes, S. M. Cohn, K. R. Corum, D. L. Greene, D. M. Hamblin, G. Kulp, D. B. Reister, T. A. Vineyard. 1981. *Energy Use from 1973 to 1980: The Role of Improved Energy Efficiency.* ORNL, CON-79 (December).

Jet Propulsion Laboratory. 1975. *Should We Have a New Engine, An Automotive Power System Evaluation* (August).

Kempton, Willett, Craig K. Harris, Joanne G. Keith and Jeffrey S. Weihl. 1982. "Do Consumers Know What Works in Energy Conservation?" Institute for Family and Child Study, Michigan State University, presented at the 1982 Summer Study American Council for an Energy-Efficient Economy (August 23) Santa Cruz, CA.

Mellon Institute of Research. 1983. *Project Evaluation Residential Energy Efficiency Program: Pilot Demonstration in Lakewood, New Jersey.* DOE/CS/22054-1 (August).

National Materials Advisory Board. 1983. *An Assessment of the Industrial Energy Conservation Program for the Pulp and Paper and General Manufacturing Industries.* NMAB 407-2. Washington, DC: National Academy Press.

________________. 1982. *Critique of Department of Energy - Office of Industrial Program Fiscal Year 1982 Research and Development Programs.* NMAB 407-1. Washington, DC: National Academy Press.

________________. 1981. *An Assessment of the Industrial Energy Conservation Program.* NMAB 395-1. Washington, DC: National Academy Press.

OTA (Office of Technology Assessment). 1982. U.S. Congress. *Energy Efficiency of Buildings in Cities* (March). Washington, DC: U.S. Government Printing Office.

Savitz, M. 1984. Testimony before U.S. Congress. House Committee on Science and Technology (June 13).

Schipper, L. and S. Meyers. 1983. "Energy in American Homes: Conservation Takes Hold." Energy Analysis Program, Lawrence Berkeley Laboratory, University of California, Berkeley, CA, LBID.708 (June).

Socolow, Robert H. 1978. *Saving Energy in the Home: Princeton's Experiment at Twin Rivers.* Cambridge, MA: Ballinger Publishing Company.

Stern, P. and E. Aronson (eds.). 1983. *Energy Use: The Human Dimension.* National Research Council, National Academy of Sciences (March). Washington, DC.

U.S. House and Senate Committee Reports. 1975-84. Department of the Interior and Related Agencies Appropriation Bills.

Commercialization Issues In Energy Technology Policy

J. David Roessner

Technological innovation is a key to accomplishing significant national energy goals. In recognition of this, government programs in industrialized nations have, for decades, attempted to stimulate energy production. Government incentives have been used to facilitate the process by which energy technologies are developed, tested, demonstrated, produced, and disseminated. Between 1950 and 1976, France, Germany, Japan, and the United States spent a total of nearly $300 billion on incentives for energy production: $477 per capita in France, $871 in Germany, $300 in Japan, and $870 in the U.S. (Cole et al., 1981). Such initiatives were expanded in both magnitude and scope after the Arab oil embargo in 1973-74 and the subsequent rapid increase in petroleum prices. Governments of many oil-importing, industrialized nations launched aggressive programs to reshape the entire pattern of national energy production and use. The central objective of these efforts was to reduce reliance on petroleum by speeding up innovation and diffusion of alternative energy options - coal, nuclear energy, renewables and conservation.

In the United States, many of the new programs were developed with the clear recognition that while government might facilitate the process of innovation, private firms must eventually produce and sell new energy technologies. Energy policies to induce technological innovation in the private economy embodied conscious strategies of commercialization. By intervening in the processes of development and demonstration, government would speed up the production and use of new technologies in the marketplace. Energy commercialization

---

Author's Note: This chapter develops and extends the ideas contained in Roessner, 1984, 1982a, and 1980b.

strategies were meant to bridge the gap between the processes of invention and those of social adoption and technological diffusion. Such efforts represented a significant extension of the government's historical role in helping to create a national climate conducive to innovation.

The U.S. Department of Energy (DOE) was given an explicit mandate by Congress to foster the commercialization of new technologies. Across the entire range of energy supply and conservation technologies, DOE has devised and implemented programs to accelerate the pace with which energy technologies will be introduced, produced, and purchased in the marketplace. Such government programs are justified by the existence of acknowledged market imperfections and failures that might otherwise prevent a socially optimum level of innovation. The perception of direct threats to national security interests posed by the dependence on imported oil also serves to rationalize more active government intervention in energy markets.

During the Carter administration, there was no serious challenge made to the appropriateness of government programs for the commercialization of energy technologies. The Reagan administration, however, has raised a number of objections to such programs and has adopted a more restrictive definition of appropriate government interventions in energy markets. Despite this, the major difference between the Carter and Reagan positions is confined to a relatively narrow issue: at what stage in a technology's development is government involvement no longer justified. Combining welfare economics principles with selected information on the existence and severity of an energy crisis, both administrations produced rationales for what amounted to ideologically-based commercialization strategies for energy technologies. Regardless of the Reagan administration's efforts to reduce the government's involvement in commercialization programs, increasing international economic competition, national security requirements, rising energy prices, and the possibility or actuality of energy shortages will continue to force policy makers to consider solutions involving government programs to accelerate energy technology development.

R&D policy, including commercialization policy, will almost certainly influence the range of available energy technologies and the rate of their diffusion. But which government policies will, in fact, best accelerate energy technology development and, more important, promote widespread adoption of any such developments? Is government intervention productive at some points in the energy technology commercialization process, but ineffective at others? What kinds of government intervention will further the process at such points? In short, what is known about the factors influencing the commercialization process, and what can we now say about the appropriateness of recent energy commercialization strategies?

Energy Policy and Technological Innovation

Federal programs designed to promote technological innovation have raised an entirely new set of public policy issues concerning the value, costs, timing, and efficiency of government involvement in the private economy. The issues posed by government efforts to influence the rate and direction of technological change are of at least three types. General policy issues involve basic questions of what the government's role should be in the process of technological innovation. Is government-supported action to influence technological change warranted? If so, how should the costs, benefits, and risks be distributed and shared among the multiple public and private actors in the innovation process? Over what period should government intervention extend? The overarching criterion to apply in answering the fundamental question of governmental involvement is that of net social benefits resulting from the action (Costello et al., 1978). How the benefits and costs of intervention are to be effectively identified and measured, however, is far from clear and a continuing subject of policy debate.

A second category of concerns relates to systemic or program level issues. Here, the government's role is assumed to be justified and attention is focused on questions of how to carry out program objectives. Systemic problems of this type involve selecting effective and efficient strategies for increasing innovative behavior. How can the overall rate of market penetration of innovations be increased? What general tools are available to government to intervene effectively in the process?

Finally, there are technology-specific issues. How can the degree of market penetration of particular technologies be increased? What factors will influence the rate of market penetration and to what extent and in what ways can these factors be affected by government programs?

All of these issues have been apparent in DOE's efforts to promote the development and application of energy technologies. They emerge most clearly in the specific policy context of particular energy technologies. DOE's programs for solar energy technologies are a good example. Such programs can be divided into two categories: those where the primary objective is to reduce the costs and improve the performance of particular energy technologies; and those where the primary objective is to provide incentives for, or reduce barriers to, the rapid spread of energy technologies. Cost reductions and performance improvements are desirable because they make the technologies more economically competitive in potential markets. Thus, for example, many solar energy programs have established technology cost goals defined as the costs that would make solar systems competitive with other energy sources. Within DOE solar energy programs, numerous technologies and approaches to reduce cost and improve performance have competed for R&D funds. Estimates of the energy contributions that can be expected from each technology and when these contributions will be realized are fundamental to the budget allocation process. For the most part, market penetration estimates of solar energy have treated cost as the only important factor (Schiffel et al., 1978). Nonetheless, a number of factors in addition to cost are known to influence the diffusion rate of new technologies: technical reliability, convenience and comfort; aesthetics and life-style conditions; and, institutional factors such as the availability of information and financing, the impact of building codes, and so on (Roessner, 1982).

Commercialization issues generally arise when technological cost and performance goals largely have been met and the technology is assumed to be ready for market introduction. Government actions are then directed toward developing strategies other than cost reduction or performance improvement that will speed the adoption and use of the technology. Demand for the technology can be increased through financial incentives; uncertainty or ignorance among potential consumers can be reduced through demonstrations, information dissemination, or product certification and warranties; institutional barriers such as resistance among lending institutions and lack of trained installation and maintenance personnel can be addressed through incentives, information dissemination, and education campaigns.

There are three separate challenges for effective commercialization policies: (1) correctly identifying the problem impeding technology diffusion or market penetration; (2) selecting the strategy or mix of strategies appropriate to the problem; and (3) designing and implementing programs that will be effective in each particular technology and market setting. As in the case of R&D programs generally, DOE budget allocation decisions for commercialization depend on estimates of the relative success of alternative strategies or strategy mixes for each technology, since the proportion of national energy produced by target technologies will depend on the extent of their market penetration.

Table 1 presents a series of policy issues confronting managers of both energy R&D and commercialization programs. Parallel to these issues are questions requiring knowledge of the rate and process of energy technology market penetration, behavioral adoption, and social diffusion. These issues and questions suggest that energy commercialization programs, as well as research and development programs in general, could benefit from the application of research on processes of innovation and technological diffusion.

Table 1
Energy Policy Issues and Related Commercialization Questions

Policy Issues	Commercialization-Related Questions
Cost Reduction and Performance Improvement	
1. If specified cost and performance goals for energy technologies are achieved, what amount of energy could each technology produce in the future?	1. What effect do cost and performance have on the commercialization of energy technologies?
2. If different research and development budget allocations are expected to result in different future technology performance and cost characteristics, which allocations will result in the most rapid and extensive commercialization?	2. What factors other than cost and performance affect the rate of commercialization?
3. How will the diffusion rate affect the success of cost reduction strategies such as government procurement, which are designed to reduce cost through increases in production rates and accumulated production experience?	3. What is the relative significance of cost, performance, and other factors on the rate of commercialization of technologies?
	4. How can the many factors that affect commercialization combined to estimate future energy market penetration?
	5. How accurately can data collected during early stages of commercialization predict the ultimate pattern and rate of commercialization?
	6. Will government and private purchases of a technology be sufficient to permit economies of scale in production?
	7. Will reductions in price lead to increases in private purchases that will cause further cost reductions attributable to accumulated industry experience?

Table 1 (Continued)	
Policy Issues	**Commercialization-Related Questions**
Barriers and Incentives	
1. What incentive strategies are most likely to be effective in accelerating the commercialization of energy technologies? 2. How should specific incentive programs such as information dissemination or demonstrations be designed and implemented? 3. What is the appropriate balance between government actions to further improve cost and performance, and actions to stimulate adoption of existing systems? 4. How is this balance to be determined on a technology-specific basis?	1. What is the relative significance of particular barriers to energy commercialization? 2. What will be the effect of alternative incentive options upon these barriers and the commercialization rate of a particular technology? 3. To what extent do selected incentives work together to simultaneously reduce barriers and increase commercialization? 4. How is information exchanged in specific market sectors? 5. What types of demonstration projects could be successful in reducing uncertainty, how many projects are needed, and where should they be located? 6. To what extent is lack of information delaying use of energy? 7. What kinds of information would best assist different groups in using energy?
Source: Adapted from Roessner, 1980.	

Models of Industrial Innovation and Technological Delivery

Private firms and institutions are key elements in technological innovation and delivery. To intervene effectively in the process by which private firms and institutions successfully market a new idea, program designers and policy makers must accurately anticipate the timing of interactions among the participants and correctly identify the factors that influence their behavior. In particular, if government programs to accelerate the pace of technological innovation in energy markets are to result in net positive benefits, those who design and administer them need to understand the typical evolutionary path of industries.

The process by which innovation-based industries originate and evolve has been studied by a number of scholars, and rough conceptual outlines of this process have been developed. The role that innovation plays in corporate strategy, the focus of internal investment, and an industry's response to government policies and incentives vary considerably as the industry evolves. Abernathy and Utterback (1978) describe a dynamic model of product and process change in industry that begins with one or more major product innovations. In the beginning stages of the evolution of an industry, the initial product line is diverse, consisting mainly of custom designs. Emphasis at this stage is on improvements in product performance, which are often the result of information from the customer's needs and technical inputs. Later, as at least one product design emerges that is stable enough to warrant a significant volume of production, attention and resources shift toward production process improvements and product variation. Finally, cost reduction through process improvement dominates investment and innovation strategies. As this process occurs, manufacturing shifts from flexible, small-scale units reliant on manual labor and craft skills to automated, equipment-intensive, high volume processes (Utterback, 1979).

Mueller and Tilton (1969) describe a different, although complementary, four-stage model of the evolution of an innovation-based industry. In the innovation stage, large commercial and technical uncertainties exist, and economies of scale in R&D, production and marketing are not yet feasible. Small firms have several advantages over large ones at this stage, and often dominate industry activity. At the imitation stage, the demonstrated success of innovating firms reduces technical and marketing uncertainties, opening the way for entry of both small and large firms. The costs of acquiring knowledge are not great, and R&D will be conducted on a trial-and-error basis using makeshift equipment. As the industry passes into the technological competition stage, the acquisition of detailed knowledge of the scientific and technological principles underlying the innovation determines the type of firms that compete successfully. Large research laboratories and small firms already in the industry that have developed an area of specialized knowledge are likely to prosper. At this stage, a number of small firms will be acquired by large ones. Entry will be difficult because of the need to obtain specialized knowledge required to produce products that sell because of their high quality or low price. In the standardization stage, competition shifts from technology to price as production techniques become standardized. Barriers to entry are mainly due to the large capital investment required to develop efficient production and marketing capabilities.

Such models of industrial development typically address only the producer function in the innovation process, ignoring most of the other elements described earlier. Yet the existence and appropriate functioning of other actors can be critical. This is acknowledged in the concept of a 'technology delivery system,' a "symbolic network that portrays technology as a potent agent of social process, symbiotically related to the institutions involved and acting in each delivery system as motivation and source of coherence" (Wenk and Kuehn, 1977; 153-65). Ezra's application of the concept of solar energy technology for building markets illustrates both the key role of institutions (such as local building codes, banks, and standards agencies) in successful innovation, and the myriad ways in which public policy can influence the speed and direction of innovation (1975).

Taken together, evolutionary models of innovation-based industries and the concept of technology delivery systems suggest strongly that public policy makers must be sensitive to the dynamics of interaction among the components of the system and to changes in the system as the industry and its underlying technology evolve. Government commercialization strategies should match the stage of evolution of the target industry and its associated technology. Table 2 illustrates what Utterback (1979) considers appropriate matches between the phase or stage of development in an innovation-based industry and related government policy actions. In his discussion of the policy implications of a developmental model, Utterback does not consider the possibility that government actions might also be appropriate if, at any stage in an industry's development, weaknesses in the technology delivery system exist. This possibility is presented in Table 3. The two tables together comprise a framework for analysis of U.S. government strategies for commercialization of energy technologies. For example, if an industry and technology are in Mueller and Tilton's "innovation" stage, government commercialization strategies should take account of the technology's weak technical base, the existence of large technical and market uncertainties, and the possibility that key elements of the delivery system are missing or weak.

Table 2
Phase or Stage of Industry Development and Corresponding Government Commercialization Strategies

Phase or stage of development		Corresponding government commercialization strategies
Abernathy and Utterback	*Mueller and Tilton*	
	Innovation	
Start-up		R&D support; loans and loan guarantees; provision of venture capital; facilitate market entry; provide adequate skills
	Imitation	
Growth		Exchange information; develop standards; invest in plant and equipment
	Technological Competition	
Stable	Standardization	Maintain competition; limit social costs of use of products and processes; ameliorate dislocations due to structural changes in the economy

Source: Based in part on Utterback, 1979.

Table 3
Element of Technology Delivery System and Corresponding Government Commercialization Strategy

Weak delivery system element	Corresponding government commercialization strategy
Producer	Loans, tax subsidies, R&D contracts
Technology	Research, development, engineering tests
Demand	Guaranteed purchases, income tax credits to buyers, demonstrations, consumer information
Infrastructure	Standards, installer/maintenance training, product testing, information, consumer protection

Source: Roessner, 1984.

While the need for "appropriate matches" may seem obvious, in fact the U.S. government's commercialization strategies for energy technologies often have not matched the evolutionary status of the industry and technology involved. If deviation from this principle is substantial, the negative consequences for the pace of technology commercialization and, indirectly, for the taxpayer might also be substantial. An examination of commercialization strategies for solar energy technologies suggests that the deviation indeed has been substantial and the consequences of this deviation costly.

Solar Energy Commercialization Strategies

Government commercialization programs for solar technologies have been justified using a combination of principles of welfare economics and the perception that an energy crisis existed. Because of existing subsidies for conventional fuels (like the oil depletion allowance and the regulation of natural gas prices), the reluctance of firms to invest in technologies with uncertain markets and/or long-term payoffs, and the existence of a variety of institutional "barriers" to the acceptance of solar technologies, a substantial government role in promoting the commercialization of solar energy has been rationalized. The nature of that role has depended on the perceived stage of development of the technology.

The outcomes of these solar commercialization strategies after five years are far from clear. The current status and past trends of U.S. solar industries and markets cannot easily be accounted for in terms of the actions of specific government programs; for example, no well-accepted method exists for determining whether government research and development funds displaced or supplemented private funds. What can now be evaluated is whether DOE's commercialization strategies for each technology were well-suited to the development stage of the target industry and the technology. I shall consider three technologies: flat plate collectors for solar heating, solar thermal concentrating collectors for process heat, and photovoltaic cells and modules. The three technologies and their associated industries represent a range of technological sophistication, markets, stages of industrial development, and stages of infrastructure development.

Active Solar Heating

In the case of solar water heating for residential applications, the technology was perceived to be "market ready." Only financial incentives to prospective purchasers to overcome distorted energy prices and a variety of programs (e.g., demonstrations, information dissemination) to reduce institutional barriers to acceptance were thought to be required. The residential and commercial demonstration programs and the solar tax credits enacted under the National Energy Act reflected this strategy. It was assumed that technological uncertainties had been reduced to a very low level. The perceived market-ready status of solar heating technology meant an emphasis on demand-oriented strategies: residential and commercial demonstrations and income tax incentives for purchases of solar systems. Through 1980, the major programs of these types were:

1. Residential and commercial solar demonstration projects, including a residential demonstration program, a federal buildings program, and a "hot water initiative." Total expenditures for these programs since 1975 were approximately $80 million (Morse, 1977).

2. Residential and business income tax credits available under the provisions of the National Energy Act and of a number of state solar financial incentive programs. Public expenditures under the NEA for solar tax credits for 1978 and 1979 totaled approximately $75 million (U.S. Internal Revenue Service, 1978; Solar Energy Intelligence Report, 1980).

The number of firms manufacturing flat plate solar collectors grew from a few dozen in 1973 to well over 300 in 1977. Since then, the number of firms has remained roughly constant at about 300. Most of the firms in the industry are small, independent ventures; more than half the collector manufacturers employ fewer than ten persons (Levy and Field, 1980).

Estimates of sales by solar collector manufacturers exhibit a rapid increase between 1974 and 1977, followed by a continuing period of slower growth. The most financially profitable element of the business was production of swimming pool heaters, where a half-dozen firms shared sales revenues of about three million dollars. The medium-temperature and special collector market was aimed largely at residential heating applications and contained a much larger number of firms competing for a market share. The low-temperature market has exhibited considerable concentration of revenues in a few firms within the last several years. A similar trend may be beginning in the medium-temperature market, where in 1982 fifteen firms shared 51 percent of the total sales (Solar Age, 1983). In 1980, however, no single firm served even 10 percent of the market (Booz, Allen and Hamilton, Inc., 1980).

Barriers to entry in the solar heating industry have been relatively low, but the exit rate has been high. The high exit rate is often attributed to inadequate technical business skills of solar entrepreneurs. Adequate returns on investment are rare, and many companies experience net losses. Industry sales seem insufficient to sustain the existing number of firms, but a "thinning out" process has yet to occur.

In 1974, Congress appropriated money to produce and disseminate information about residential solar systems and to demonstrate the technology to the construction industry. The judgment that residential solar heating technology was ready for the marketplace was made in the context of advocates' claims of commercial readiness and the absence of visible opposing information. Commercialization efforts focused on market demand through subsidies to purchasers, yet solar heating technology was, in the mid to late 1970s, still characterized by substantial technical uncertainties (U.S. General Accounting Office, 1980; Koontz, Genest and Bryant, 1980). Despite this, programs oriented toward strengthening supply (through loans and loan guarantees to small businesses) and infrastructure (through training programs for installers and consumer protection) received relatively little federal attention and support (U.S. Department of Housing and Urban Development, 1980).

A documented record of technical failures and disappointing performance of installed residential solar systems has revealed a mixture of faulty design and improper installation (U.S. Department of Housing and Urban Development, 1980; U.S. General Accounting Office, 1980; Koontz, Genest and Bryant, 1980). A detailed study of the performance of 60 installations in Florida showed that many were improperly installed and/or maintained, suggesting weaknesses in installer training, system monitoring techniques, and maintenance. The authors of an article reviewing the Florida findings concluded that "gross inadequacies [existed] in the solar energy delivery system" (Yarosh and Connor, 1980).

The solar heating industry in 1975-80 exhibited a weak technical base, poorly developed infrastructure, and inadequate management and marketing skills. It did not possess the resources to train and manage installers, quickly identify and correct technical problems with system components, or reduce system costs and improve reliability through economies of scale or R&D. The industry clearly was in the innovation stage when the DOE strategy was formulated. Since substantial uncertainties about the technology's performance and reliability remained, the "commercially ready" label applied by the government was inappropriate at the outset.

Demand-oriented commercialization strategies, such as market demonstrations and financial incentives to potential purchasers, assume that the technology works and works well, and that key elements of the technology delivery system are in place and functioning well. In the case of solar heating, these assumptions did not hold. A strategy more appropriate to a competitive or mature industry was applied to an emerging one. A more carefully chosen strategy would have recognized the systemic character of weaknesses in the solar heating industry and allocated resources to supply and infrastructure elements of the delivery system in amounts that balanced those devoted to stimulating demand.

Concentrating Collectors for Solar Thermal Applications

DOE's solar thermal program had two major objectives: to reduce system costs to approach commercial competitiveness by the mid-1980s; and to demonstrate "acceptable" reliability, environmental characteristics, and maintenance costs (U.S. Congress, 1979: 254). The program supported R&D to reduce collector costs and improve collector ability to handle high temperatures, and obtain utility and industry experience through "field experiments." Specific cost goals, in 1979, were to reduce small system industrial and agricultural applications costs to twice market value by 1983. Twenty to thirty percent of the total DOE solar thermal program budget during 1976-80 was devoted to market tests and demonstrations focused on such applications; more than $100 million was spent for construction or capital equipment purchases and for market tests and applications (Braun, 1979). The bulk of DOE expenditures during that period, however, was for R&D and engineering tests of various concentrator designs, materials, and configurations. There was very little documented private R&D support for related activities.

Until the Reagan administration took office, the concentrating collector industry was supported almost entirely by government purchases for tests and experiments. Based on DOE project costs, collector manufacturers may have received a total of about $30 million as of 1980 for the sale of collectors (Roessner, 1980). With only a few exceptions, capital investment in the concentrating collector industry has been supported by the government. Only four installations were privately financed prior to 1980 (Sargent, Glenn and Kearney, 1980).

Only about 20 firms manufacture concentrating collectors designed for industrial process heat and electric power applications. These firms are diverse, including small private ventures, small divisions of large companies engaged in related activities, and divisions of large energy, electronics, and aerospace conglomerates. Large architectural and engineering firms have been involved in system design and construction because of the scale of many installations requiring tens of thousands of square feet.

The situation in 1980 suggested that greatly reduced uncertainty about future markets for concentrating collectors was necessary before any private firm would invest substantial capital in production facilities. This would occur only through increasing knowledge about the performance and reliability of solar installations in a wide range of industrial and utility settings; these installations and their monitoring apparently had to be supported with public funds. While some members of Congress called for financial incentives for solar industrial applications (a 50 percent tax credit and a 7-year depreciation period for capital investment), the effectiveness of these incentives is questionable, given the technical uncertainties that remained. Industry representatives at an industry-government forum in 1980 were agreed that the key to solar thermal market penetration was user acceptance, which could not be attained through economic incentives for an unproven technology (Science Applications, Inc., 1980). Industry was concerned that technical options would be narrowed prematurely in the DOE program. In 1979, the program's chief expressed his concern that, under pressure from Congress and the Office of Management and Budget to narrow options, set priorities, and allocate funding to the most promising technical options, critical technical issues in several alternative subsystem configurations might go unresolved (Braun, 1979). In addition, many observers believed that the DOE program supported field tests and demonstration projects prematurely, before the technology was ready. The poor operating performance of concentrating collectors in field demonstrations suggests that this assessment is accurate (Lumsdaine, 1981; Best, 1982). As one critic of DOE's industrial process heat program put it, "if the Department of Energy's intention was to find out what might go wrong with industrial solar energy programs in the field, then the program succeeded spectacularly" (Best, 1982: 29).

The concentrating collector industry certainly is a newly emerging one, but it does not resemble Mueller and Tilton's characterization of the innovation stage. Understandably, their model does not fit a situation in which an "industry" is created virtually overnight, whose profits are based on overhead and fees from government R&D contracts, and whose sales are almost exclusively to government. Government commercialization strategies for concentrating collectors reflected no single view of the industry: substantial R&D spending and engineering tests suggested government perceptions of an industry and technology in the innovation stage; equally substantial investments in demand-oriented strategies suggested perceptions of an industry and technology evolved to a later stage. To the extent that government programs emphasized R&D to reduce system costs and improve performance and reliability, they were appropriate to the technology's status. The industry apparently was capable of absorbing and using government R&D funds effectively.

Difficulties arose when demand-oriented strategies became a significant element of the process heat element of DOE's solar thermal program. What were labeled as field experiments were regarded, appropriately or not, as demonstrations. To the extent that these experiments/demonstrations revealed substantial weaknesses in the technology's performance to potential purchasers, they were premature. At the close of the Carter administration, there was no consensus on the government's role in the concentrating collector industry and very little analytical work to illuminate the question. The Reagan administration's philosophy resolves the question clearly in the direction of minimal government role.

Photovoltaics

From its inception through 1980, the major objective of DOE's photovoltaic program was to reduce the cost of installed photovoltaic systems to the point where, by 1990, costs of grid-connected residential systems would be competitive with conventional energy sources. These cost reductions were to be achieved by "aggressively pursuing advanced research and technology development" (U.S. Congress, 1979: 265). Evidently, the relatively long history of photovoltaic development had reduced technical uncertainties to a low level; cost competitiveness was to be achieved via technological change resulting from R&D. Since cost competitiveness required that photovoltaic costs be reduced by a factor of 100 or more, subsidies to purchasers were not feasible.

Cumulative expenditure for the U.S. photovoltaic program totaled nearly $450 million for the period 1975-80. The program emphasized advanced research and development ($95 million in FY 1977-80), technology development ($163 million in FY 1977-80), and test and applications ($99 million in FY 1977-80). The program employed a "supply" approach to commercialization: most program expenditures were for R&D directed toward system cost reduction. Beginning in 1978, however, Congress passed several bills apparently intended to increase the demand-oriented (or purchaser-oriented) activities of the photovoltaic program, including a $13 million program for the purchase of photovoltaic systems for installations and a $1.5 billion program over a 10-year period that emphasized applications rather than R&D. These programs were not implemented extensively and, under the Reagan philosophy, will not be revived in the foreseeable future.

The photovoltaic industry in the United States consists of two groups of firms, only slightly overlapping, that engage in very different types of activities, employ different investment, marketing and operating strategies, and hold different views about the appropriate federal role in the industry. One group of about 15 companies currently manufactures photovoltaic cells and/or panels for sale in the market. Although most companies that produce photovoltaic cells also conduct some R&D, the latter activity does not define their current role in the marketplace. The current manufacturers of photovoltaic cells or modules include independent companies, subsidiaries of oil companies, and subsidiaries of electronics firms. The second group, numbering over 80, engages primarily in research and development related to photovoltaic production process development or materials. During 1980 and 1981 several companies shifted some of their activities from R&D to production: among these were Photon Power, ARCO Solar, and Mobil Tyco (Roessner, 1982).

The photovoltaic industry has been described as "growing, healthy, and competitive" (Jet Propulsion Laboratory, 1980). A basic core of cell and module producers has remained in business for the past five years, with only a small number of exits and a moderate number of new firms appearing. Capital requirements have been met at least moderately successfully through acquisition and new equity issues. Although public relations play a role in the decisions of major, diversified firms to acquire small photovoltaic businesses or to initiate a new venture, both near- and long-term investment opportunities are also important. Expected growth in foreign markets for photovoltaics in the early to mid-1980s has prompted a number of linkages between U.S. photovoltaic companies and large, diversified European firms.

Sales revenues of cell and module producers more than quintupled between 1976 and 1980, growing from more than $6 million to an estimated $39 million in 1980 (Table 4). A recent estimate by DOE (which has a tendency to use optimistic sales figures) placed 1982 U.S. sales at $100 million (Prince and Barrett, 1982: 26). Although there is little agreement on the amount of private investment that has gone into the development of photovoltaics, the figure is substantial, probably more than $100 million (Roessner, 1982).

Despite some flirtation with demand-oriented stimuli, the DOE photovoltaic program emphasized R&D and engineering tests. The program strategy itself was regarded as relatively stable until the change in administrations, although there was resentment in the industry during the Carter administration toward rapidly changing departmental policies resulting from frequent congressional and presidential involvement. The reactions of industry executives to the federal program were described in a 1978 Booz-Allen study and were probably still valid through 1981. Some companies (about one-third of those contacted) thought that the program was a positive influence on their business base and on advanced materials research and development. Nearly half the companies thought that the overall effect was negative for a variety of reasons, including public discussion of prices, confusion of price with cost in the public mind, unclear goals, premature commercialization of the technology, and lack of program and funding continuity. The remaining firms (about 20 percent) did not think that the government program had any significant effect on the industry.

No simple definition of the development stage of the photovoltaics industry suffices. Current producers of cells and modules seem to have entered the technological competition stage based on a relatively well-established technology, the single-crystal silicon cell. Firms investing in R&D in order to achieve technological breakthroughs would more properly regard themselves as part of an industry at the innovation stage of development. (Again, the limitations of simple evolutionary models are apparent.) The DOE photovoltaic program devoted some of its resources to block purchases and demonstrations (demand-oriented strategies), but most went to R&D programs having dramatically reduced cost as their goal. The second, third, and fourth generation technologies that were, and still are, the focus of the program represent state-of-the-art materials research. This dual strategy generated some tension. Current producers of cells and modules were displeased because an emphasis on R&D implied that the technology was not commercially ready; R&D also shortens the time when current technology will become obsolete. In contrast, R&D performers believed that few public resources should be spent on current technology. DOE's photovoltaic program was implemented against a backdrop of substantial private investment in the technology, particularly R&D investment. As a supplement to private efforts rather than a driving force, this supply-oriented strategy created few problems other than uncertainty resulting from the inevitable vagaries of public policy. As in the case of concentrating solar thermal collectors, greater difficulties arose as massive, demand-oriented strategies were proposed by the Congress and contemplated by DOE.

Through 1980, however, the DOE program appeared well-matched to the status of photovoltaic technology and the photovoltaic industry. Substantial cost reductions in the technology were observed over the period, and for the most part, pressures to shift toward demand-oriented strategies were resisted successfully. There was no evidence of significant breakdowns or malfunctions of the experimental installations, and government R&D investments were supplemented by substantial private investment in both R&D and production facilities.

Under the Reagan administration, the DOE photovoltaic program has undergone substantial reduction. Industry reactions are mixed, with some representatives glad that photovoltaics will "stand on its own" and others fearing increased competition from European and Japanese firms subsidized by their governments. In comparing the current U.S. and Japanese photovoltaic programs, the former DOE program manager laments a move toward "funding research for the sake of research" (Maycock, 1982: 15). In his judgment, the balance and increasing funding of the Japanese program are offset to some extent by the fact that the Japanese offer no tax credits to purchasers or buy-back guarantees for small photovoltaic generators who connect to the grid. But, on the whole, he is not optimistic about the ability of the U.S. to retain its share of the world market (Solar Age, 1983).

Conclusion

Evolutionary models of innovation-based industries and the concept of technology delivery systems provide a useful framework within which to assess the effects of government energy commercialization strategies. As demonstrated by review of the three solar energy industries, however, such a framework is still too crude to offer detailed predictive or explanatory power. Furthermore, the industrial innovation models were derived from, and seem most applicable to, industries originating with a technologically sophisticated innovation such as the transistor or microprocessor. Not all important energy innovations will exhibit these characteristics. If the active solar heating industry remains decentralized in both production and distribution, as seems likely, it will not follow the predicted evolutionary trend toward mass production and capital intensiveness. The concentrating collector industry, essentially a government creation, is just beginning its evolution. There is no single innovation, technically sophisticated or otherwise, that characterizes the industry. Major concentrating collector producers include aerospace companies and large conglomerates, and if the industry develops, they will continue to play a major role. The photovoltaics industry is bifurcated, but several R&D performers are becoming producers as well. Of the three industries examined here, the photovoltaic industry's development most closely approximates the evolutionary models' predictions. The R&D-intensive nature of photovoltaic cell production and of continuing developments in the basic technology partially account for this relatively close match with the models. The characteristics of technology delivery systems have been frequently neglected in industrial innovation research, but the review of solar technologies indicates that these characteristics may be critical to successful commercialization.

No single explanation accounts for the mismatches observed between the status of the three solar energy technologies and the strategies used by the federal government to stimulate their development. What is clear is that energy commercialization strategies need to be better understood as products of the prevailing politics of technology policy. Thus, for example, the reward system of public policy makers helps explain the overall timing and design of U.S. solar technology policy. In the crisis atmosphere following the Arab oil embargo, members of Congress were eager to show their constituents that they were attacking the problem. Political benefits were reaped from highly visible, easily implementable, short-term programs that touched as many different constituencies as possible. The pressure to show results probably accounts in part for the premature emphasis on large-scale demonstrations in residential solar heating and in industrial process heat. Solar tax credits, which were implemented at about the same time as residential and commercial demonstration programs, met the necessary political criteria. Insofar as the DOE photovoltaics program began in 1979 to plan for large-scale, government-subsidized purchases of photovoltaics, it, too, was responding to these kinds of pressures from Congress. The relative success of the photovoltaic program in resisting such pressure could be a function of the DOE program manager's experience and skills, the acknowledged large gap between current photovoltaic prices and the price of competitive energy sources, the quality of DOE program planning in photovoltaics, or the political strength of photovoltaic firms that emphasized R&D. These political and administrative influences on the direction of commercialization strategies are not unique to solar technologies, and frequently such influences overwhelm more analytic evaluations of appropriate commercialization initiatives.

Case studies of the U.S. government's involvement in industrial innovation (Nelson, 1982) reveal a pattern of relatively successful intervention in basic research, "generic" technologies, and fundamental research areas such as health and agriculture where researchers, rather than government officials, make resource allocation decisions. When government attempted to "pick winners" and to intervene in the later stages of technological development, the results were substantially less favorable. Such findings appear consistent with experience in energy commercialization. Two sets of factors help to explain these outcomes. On the one hand, government officials often have insufficient knowledge of civilian markets and technology delivery systems, and that may well have been the case in the renewable energy programs of the U.S. Department of Energy. On the other hand, the political and administrative dimensions of technology policy often result in commercialization efforts inappropriate to the state of the technology or to the market and institutional conditions available for its delivery. The interactive role of these factors in technology policy formulation and implementation are still not well documented and understood. As a result, energy commercialization strategies may frequently have been less successful than they might have been.

REFERENCES

Abernathy, J. and M. Utterback. 1978. "Patterns of Industrial Innovation." *Technology Review* 80 (June/July): 41-47.

Best, D. 1982. "Industrial Process Heat: Taking the Private Plunge." *Solar Age* (February): 28-31.

Booz, Allen and Hamilton, Inc. 1980. *Characteristics of the Solar Heating and Cooling Equipment Industry*. Bathesda, MD: Booz, Allen and Hamilton.

Braun, G. 1979. DOE Solar Thermal Power Systems Program. In *Solar Industrial Process Heat Conference Proceedings. Golden, CO:* Solar Energy Research Institute.

Cole, R. J., B. W. Somers, P. Eschback, C. Sheppard, W. J. Lenerz, D. E. Huelshoff and A. A. Marcus. 1981. *A Comparison of the Incentives Used to Stimulate Energy Production in Japan, France,*

West Germany and the United States (PNL-3788). Richland, WA: Battelle Pacific Northwest Laboratory.

Costello, D. et al. 1978. "Photovoltaic Venture Analysis, final report, Volume 1." Golden, CO: Solar Energy Research Institute.

Ezra, A. A. 1975. "Technology Utilization: Incentives and Solar Energy." *Science* 187 (28 February): 707-13.

Jet Propulsion Laboratory. 1980. *Federal Policies to Promote the Widespread Utilization of Photovoltaic Systems.* Washington, DC: U.S. Department of Energy, Assistant Secretary for Conservation and Solar Energy.

Koontz, R., M. Genest and B. Bryant. 1980. *The National Commercial Solar Heating and Cooling Demonstration: Purposes, Program Activities, and Implications for Future Programs* (May). Golden, CO: Solar Energy Research Institute.

Levy, G. W. and J. Field. 1980. *Solar Energy Employment Requirements 1978-1985.* Columbus, OH: Battelle Columbus Laboratories.

Lumsdaine, E. 1981. "Solar Industrial Process Heat (SIPH)." In *Proceedings of the Solar Technology Assessment Conference.* Orlando, FL: Florida Solar Energy Center.

Maycock, P. D. 1982. "Fact-Finding Tour of Japan." *Solar Age* (September): 14-15.

Morse, F. B. 1977. "Status of the U.S. Heating and Cooling Program." Washington, DC: U.S. Department of Energy, Office of Solar Applications.

Mueller, D. C. and J. E. Tilton. 1969. "Research and Development Costs as a Barrier to Entry." *Canadian Journal of Economics* 2 (November): 570-79.

Nelson, R. R. (ed.). 1982. *Government and Technical Progress.* New York, NY: Pergamon.

Prince, M. B. and A. L. Barrett, Jr. 1982. "United States Federal Photovoltaic Program Status." In W. H. Bloss and G. Grass (eds.), *Photovoltaic Solar Energy Conference.* Boston, MA: D. Reidel.

Roessner, J. David. 1984. "Commercializing Solar Technology: The Government Role." *Research Policy* 13: 235-46.

_______________. 1982a. "Government-Industry Relationships in Technology Commercialization: The Case of Photovoltaics." *Solar Cells* 5: 101-34.

_______________. 1982b. "U.S. Government Solar Policy: Appropriate Roles for Nonfinancial Incentives." *Policy Sciences* 5: 3-21.

_______________. 1980a. *An Assessment of Selected Solar Energy Industry Activities.* Prepared for the U.S. National Science

Foundation. School of Social Sciences, Georgia Institute of Technology.

_______________. 1980b. "Technological Diffusion Research and National Policy Issues." *Knowledge: Creation, Diffusion, Utilization* 2/2 (December): 179-201.

Sargent, S., B. Glenn, and D. Kearney. 1980. "Solar Industrial Process Heat." *Environmental Science and Technology* 14 (May): 518-22.

Schiffel, D. et al. 1978. "The Market Penetration of Solar Energy: A Model Review Workshop." Golden, CO: Solar Energy Research Institute.

Science Applications, Inc. 1980. *Federal Regulations Affecting Industrial Application of Solar Thermal Technology.* McLean, VA: Science Applications, Inc.

Solar Age. 1983. (July): 73.

Solar Age. 1983. (February): 44.

Solar Energy Intelligence Report. 1980. (June): 236.

U.S. Congress, Energy and Water Development Appropriations for 1980. 1979. Hearings before a Subcommittee of the Committee on Appropriations, U.S. House of Representatives, 96th Congress, 1st Session: 254.

U.S. Department of Housing and Urban Development. 1980. *Interim Report on Performance Data from the Residential Solar Demonstration Program.* Washington, DC: Department of Housing and Urban Development.

U.S. General Accounting Office. 1980. EMB-80-41.

U.S. Internal Revenue Service. 1980. *1978 Statistics of Income, Individual Income Tax Returns.* Publication 198 (4-80). Washington, DC: U.S. Government Printing Office.

Utterback, J. M. 1979. "The Dynamics of Product and Process Innovation." In C. T. Hill and J. M. Utterback (eds.), *Technological Innovation for a Dynamic Economy.* New York, NY: Pergamon.

Wenk, E., Jr. and T. J. Kuehn. 1977. "Interinstitutional Networks in Technological Delivery Systems." Pp. 153-65 in J. Haberer (ed.), *Science and Technology Policy.* Lexington, MA: D. C. Heath.

Yarosh, M. and L. Connor. 1980. "Consumer Experience: Implications for Government Solar Policy." In G. E. Franta and B. H. Glenn (eds.), *Proceedings of the 1980 Annual Meeting, American Section of the International Solar Energy Society, Inc.* American Section, International Solar Energy Society.

In Search of the Abundant Energy Machine

John Byrne and Daniel Rich

U.S. energy research and development policy has been remarkably consistent. While the political economy of energy has changed dramatically in recent decades and government energy programs have experienced significant budgetary fluctuations, national R&D policy has exhibited a nearly uniform technological preference. This consistency has resulted from the convergence of specific values and institutional conditions that have provided the framework for energy R&D policy over the past four decades. Among these, three have been of special significance. First, energy R&D has been inspired by and has sought to emulate the government's programs for weapons development during and after World War II. Second, the objective of R&D policy has been energy abundance rather than energy efficiency. Third, the evaluation of energy options and R&D allocations has been confined by a self-sealing logic of technical determinism.

Together, these values and institutional conditions constitute the normative architecture for U.S. energy R&D policy. That policy may be summarized as a search for an abundant energy machine: one which would sustain the availability of cheap energy supplies delivered by large-scale centralized institutions in a manner responsive to the demands of a high-consumption society. The prototype has been and remains nuclear power.

The Triumph of Technoscience

National R&D policy is a mid-twentieth century phenomenon. It was born out of wartime experience with weapons technology and nurtured by the postwar infusion of government resources in support of scientific and technological programs. Between the start of World

War II and the mid-1960s, the federal government spent nearly $175 billion on research and development activities. It was an event, Harvey Sapolsky points out, that was frequently taken to signal the passage of a major historical watershed. "The event's significance can be most closely compared to that of the Enclosure Acts, which transformed England into the first industrialized nation. The United States, because of the structural consequences of the government's conscious promotion of scientific and technological progress, is seen as moving beyond the societal stage of industrialization" (Sapolsky, 1975: 80). The precise nature of this transformation has been widely debated. A substantial portion of the science policy community has seemed convinced, however, that government activism in support of research and development is the leading indicator of a new era of scientific and technological prominence in social affairs. In this regard, Sanford Lakoff has pointed to the gradual evolution towards a Scientific Society: a society in which scientific understanding is both a principal measure and a key stimulus of progress. Science would become recognized as the "endless frontier," the "new property," offering incentives and opportunities to rival the stimulus historically afforded by open spaces and expanding markets. Scientists would emerge as an influential and enlightened new estate in the shaping of public affairs. Government would accept its responsibility as the patron of a scientific culture. In sum, knowledge would become power (Lakoff, 1964: 55-58; Bush, 1945; Price, 1965).

The vision of an emerging Scientific Society has offered a convenient frame of reference for interpreting the government's increased role in research and development. It does not, however, adequately identify the values and expectations that have shaped R&D policy over the past four decades. The investment of hundreds of billions of dollars in research and development is less indicative of an admiration and respect for science as a source of knowledge than an enthusiasm for large-scale technology as an instrument of national power and a confidence that with sufficient will and ample financial and organizational resources such technology can be delivered on demand. "The key intellectual commitment" according to Richard Nelson, "is the focus on a particular kind of resource-manpower with advanced formal training in the basic and applied sciences, and its employment in a particular kind of activity - organized research or development consciously aimed at achieving certain kinds of knowledge or technological capabilities" (1977: 58). Without doubt, the boundaries between basic and applied research and the interrelationship between science and technology are ambiguous and have been continuously debated. What is clear, however, is that the impetus and intent of government R&D policy has largely been to develop technologies to achieve specific national goals rather than provide general support for science. The guiding vision has been one of technological instrumentality on a grand scale: a search for the right machine.

The orientation towards research and development that arose out of the federal weapons research programs during World War II has not changed substantially over the succeeding four decades. The wartime experience, particularly the Manhattan project, established the policy link between public investment in R&D and the maintenance of national security. The vast majority of postwar investments in R&D have been intended to maintain and extend that policy link. In effect, R&D policy has been an adjunct of national security policy. Thus, Sapolsky points out that 85 percent of federal support for research and development activities in the quarter century since the war were devoted to programs to meet an external challenge (1975: 85). This condition and its implications are nowhere more apparent than in the case of energy research and development. The government role in energy R&D arose directly out of military interests, was expanded in response to the Cold War, gained urgency over the last decade to combat a perceived foreign challenge to reliable and cheap energy supplies, and continues to be conceived as a policy to protect American interests in a period of "global insecurity" (Yergin and Hildebrand, 1983).

In the decades following World War II, energy R&D was nuclear R&D. As Hewlett and Dierenfield point out, "the federal role in developing energy technologies other than nuclear in the 1950s and 1960s was minuscule" (1983: xii). The military significance of nuclear energy research and development has meant that it has enjoyed a substantial portion of the total federal R&D budget with, until recently, little necessity of substantive justification. Henry Lambright notes that at "the outset of the 1970s, 86 percent of all federal energy R&D funds that had been spent since World War II had gone to [the Atomic Energy Commission]" (1976: 33). Massive investment in nuclear technologies took place during decades when, as Nelson notes, there was no perception of energy crisis, or even of impending hard times regarding the price and availability of conventional fuel"; "the 1965 study *Energy R&D and National Progress,* which was undertaken with the express purpose of identifying sources of concern, reached the conclusion that significant shortages of conventional fuels, or sharply rising costs of extraction, were not likely in this century" (1977: 116). After the oil embargo of 1973-74 abruptly changed the energy forecast, support for nuclear research and development continued to dominate the government's energy R&D budget. Such dominance cannot be explained by nuclear's contribution to meeting U.S. energy needs now or in the foreseeable future. Indeed, firewood supplied more energy in 1980 than all civilian nuclear sources (Guerra, 1981: 20). Over the past five years, nuclear's contributions as an energy source have remained minor. Despite this, nuclear technologies continue to receive by far the largest share of government support for energy R&D. Approximately 70 percent of the Department of Energy's budget is now devoted to nuclear programs, with 40 percent dedicated to military applications (Congressional Budget Office, 1984: 31-40). This fact constitutes dramatic evidence of the military mindset that has continued to shape U.S. R&D policy in general and energy policy in particular. To paraphrase Sapolsky, the direction of energy R&D policy might best be labeled "the ballistic missile solution" to the energy crisis.

The Manhattan Project set the pattern for energy R&D programs in the decades ahead. Christopher Hill has suggested that one manifestation of that pattern has been the enormous influence granted physicists over the direction of government-sponsored R&D because of their successes in the development of weapons technologies. One result has been to promote the physicists' vision of "what constitutes good research as the touchstone of quality in other fields" and to encourage the perpetuation of high energy physics as an almost singular model for effective R&D (1985: 11). Experience in a domain of research and development which is certainly distinctive, if not unique, and which is most notable for its development of nuclear weapons, has emerged and persisted as the standard for determining the suitable subject matter and styles of federal R&D support. Energy R&D has always been dominated by the direct inheritance of nuclear power research and nuclear scientists. Even non-nuclear R&D has been characterized by periodic efforts to develop the right machine in the image of nuclear technologies (Rhodes, 1983). This situation has also resulted in a peculiar but common fascination with technological exotica and a parallel proclivity to regard the exotic as practical. As Barry Commoner has pointed out, there has been a curious inversion of expectations that has frequently accompanied the evaluation of energy alternatives in the R&D community. While solar energy, "the oldest energy source exploited by human society...was usually regarded as an impractical, exotic product," nuclear power, which is "the product of advanced science, and certainly an exotic way to boil water...was regarded as an unchallengeable, practical reality" (1979: 32).

While wartime experience served to establish nuclear physics as the substantive model for government R&D, its most significant influence lay elsewhere. As Don Price has argued, the most important discovery of wartime R&D was not scientific nor even technological in the sense of the actual "technical secrets" involved in radar and the atomic bomb. Rather, the key achievements were organizational, managerial and technocratic: the central lessons to be learned were in "the administrative system and set of operating policies that produced such technological feats" (quoted in Lambright, 1976: 23). What was seen as possible and replicable was the mobilization of manpower and money for the successful management of technological innovation on a scale that had few historical precedents. The chief lesson was that big achievements require big organizations to develop big technologies. The success of government R&D was thus seen as tied to technological and institutional scale.

Energy R&D has been dominated by large-scale programs that offer the promise of technological spectaculars, that involve billions of dollars in expenditures and casts of thousands, and that can only be managed effectively by large techno-bureaucracies, or what Lambright calls the "technoscience agencies" (1976). The institutional and technical dimensions of these scale conditions are reinforcing. On the one hand, the organizational manpower and funding patterns of large-scale R&D institutions can only be successfully accommodated to technological programs on a comparably monumental scale. On the other hand, large-scale technology programs would overwhelm the capacities of all but the largest "technoscience" institutions. Thus, for example, in the wake of the Arab oil embargo, arguments in support of the creation of a consolidated energy agency were predicated on the belief that only an overarching institution with ample and sustained funding could provide the direction for significant progress in energy research and development. The commonly identified models for such an institution were the established, mission-oriented technoscience agencies - the Atomic Energy Commission, NASA, and the Department of Defense.

Organizational and technological scale requirements have led to the rise of R&D systems whose sheer size and weight create institutional inertia. Such inertia determines the substantive focus of research and development and that focus in turn becomes progressively impervious to change. As Lambright puts it, "it is much easier to talk about redirecting R&D than it is to accomplish such a feat" (1976: 165). The nuclear program developed by the AEC is a prototype of a grand-scale technology program promoted and managed by a set of institutional arrangements that are insulated from review or change. In 1974, the bulk of this system was incorporated in the new Energy Research and Development Administration (ERDA). Part of the rationale was for this incorporation was the expectation that the well-established AEC infrastructure would provide the new agency with the scientific and technical talent, facilities and network of relationships to carry out its mission. As Hewlett and Dierenfield point out, ERDA inherited "the national system of government laboratories, university and private research institutions, engineering firms, equipment manufacturers, and electric utilities both public and private...[as well as]...the contract forms and other arrangements which the commission had devised over the years to define the role of the federal government and the place of industry in nuclear research and development" (1983: 52). Predictably, the scale and inertia of the AEC legacy largely dictated the direction of subsequent energy R&D policy. Built into ERDA and subsequently DOE was the AEC's institutionalized commitment to what constituted the right machine and how it was to be obtained. As Gregory Daneke has characterized it, "[e]nergy organization began as a tangent to the nuclear weapons development following the Second World War, and weapons remain DOE's stock and trade" (1984: 311). Advocates of alternatives to nuclear energy have found themselves overwhelmed by the magnitude of the nuclear program and the strength of its institutional support. At the same time, the institutional and technological characteristics of the nuclear program shaped the ways in which any alternative R&D programs were to be pursued.

The pattern of funding over the past decade clearly manifests this bias in energy R&D. Support for nuclear technologies continued to dominate the R&D budget. Programs for nonnuclear alternatives, such as the solar power tower experiments and the Synfuels program, were often fashioned to replicate the organizational scale and technological characteristics of nuclear R&D. Throughout the decade, support for R&D on energy efficiency options was dwarfed by investments in technologies designed to increase energy supplies. Even in the period of the late 1970s when the Carter administration and the Congress increased support for conservation and solar options, the United States was, as David Morell points out, "still basically pursuing a fuels supply policy, not one focused on energy conservation" (1981: 10). Over the past five years even rhetorical support for alternatives to large-scale production-oriented technologies has abated.

Energy R&D policy epitomizes government as a nonlearning system (Schon, 1971). The content and implications of policy have been systematically shielded from public evaluation. For decades, the commitment to nuclear technologies was officially exempted from open scrutiny. The magnitude and stability of this largely uncontested commitment resulted in a mobilization of technological bias that gradually became institutionalized as the most distinctive feature of the energy R&D policy environment. As a result, energy R&D policy has been to a significant extent a derivative policy; a by-product of institutional and technological inertia promoted and protected on grounds of national security and adapted to the needs of both the technoscience agencies and large-scale centralized systems of energy production and distribution. In this respect, the overall direction of energy R&D policy has been sustained by a pattern of nondecision making that is rationalized by the purported costs of any deviation from the current path.

Energy Myths and Megapower

The search for the abundant energy machine has been reinforced by a deeply rooted belief that energy consumption is an essential facet of social progress. As Amory Lovins has argued, underlying the contemporary evaluation of energy alternatives is the belief that "the more energy we use, the better off we are" (1977: 4). The source of this belief may be traced to what George Basalla calls the energy-civilization equation: a prevalent if often implicit assumption in Western societies over the past two hundred years that high and growing energy consumption is essential not only for physical comfort, economic well-being, and military strength, but for social, moral and intellectual achievement as well - in short, for the progress of civilization (1980: 40). In the U.S., the energy-civilization equation has seemingly been validated by recent historical experience in which growth in national wealth and energy consumption were apparently synchronous. From the vantage point of such experience, Barbara Ward has noted, a nation's wealth and its energy use are frequently two ways of saying the same thing (1977).

In the context of the energy-civilization equation, it is easy to understand why the energy events of the 1970s were seen to pose a crisis. It is also possible to recognize why policy proposals for energy conservation have frequently been greeted with skepticism. As Morell has pointed out, supply-oriented energy policies have been seen as "compatible with basic American myths and values, while conservation is seen as antithetical" (1981: 12). Reduced energy use, Basalla explains, generates apocalyptic visions of an American society in a downward slide from progress.

> We foresee a doomed civilization with tractors paralyzed in the fields, abandoned automobiles rusting on weed-choked freeways, factories as quiet as tombs, and our haggard descendants facing a life of everlasting drudgery...A retreat from rising energy consumption under those circumstances means far more than the minor discomfort of living in a warmer house in the summer and a cooler one in the winter, or driving a smaller car less frequently and more slowly. As less energy is available per capita the nation is thought to lose its standing among the world's civilizations (1980: 39,40).

Portrayals of conservation as nothing less than the progenitor of a New Dark Age have been moderated in recent years. It is no longer taken for granted that reduced energy consumption portends rapid economic collapse. Nonetheless, conservation continues to be undervalued as an energy alternative. The persistence of this condition, despite ample evidence of conservation's viability as an energy option, can be regarded as a measure of the strength of the energy-civilization myth. Precisely because it is a myth, it is largely impervious to the facts. It hardly matters that North Americans use nearly twice as much energy as the not exactly impoverished West Germans, and they in turn, about twice as much as the far from destitute New Zealanders (Ward, 1977); or that conservation has been responsible for meeting 70 percent of the new energy needs of the U.S. since the embargo (Lovins, 1983); or even that projections indicate that conservation is our least-cost supply alternative to the end of this century with a highly competitive price of $7 to $8 per barrel of oil equivalent saved (California Energy Commission, 1981). U.S. energy policies have consistently treated conservation as capable of, at best, a short-term, transient contribution along the main path to a new era of energy abundance delivered by a combination of market adjustments and large-scale technological breakthroughs. Thus, the 1983 and 1984 Reagan administration's budget proposals called for eliminating virtually all support for energy conservation R&D, while doubling funding for nuclear reactor research (Shapley et al., 1983). In the search for the abundant energy machine, conservation has been no match for megapower.

The energy-civilization equation is not the only myth that has guided our understanding of energy prospects and possibilities. Indeed, mythology has pervaded our expectations about energy. Since the industrial revolution, energy has frequently been portrayed as the essential catalyst of social organization and progress with, at times, heroic if not spiritual attributes. Odum and Odum, for example, proclaim energy as "the source and control of all things, all value and all the actions of human beings and nature" (1976: 1). Seen in this way, it is not surprising that energy sources have conventionally been invested with mythical qualities such that each new energy option has historically been characterized by a cycle of utopian aspirations followed by dashed hopes leading to a new generation of proposed energy utopias. As Basalla points out:

> [a]ny newly discovered source of energy is assumed without faults, infinitely abundant, and to have the potential to affect utopian changes in society. These myths persist until a new energy source is developed to the point that its drawbacks become apparent and the failure to establish a utopian society must be reluctantly admitted...At this point, one might suppose that society would emerge from its difficulties sadder but wiser and treat the next new energy source differently. Unfortunately this is not the case. The latest form of energy is not handled in a more restrained fashion. Instead, the recently discarded energy myths are resurrected and bestowed upon the newcomer. And so the cycle continues, with only the myths remaining unaltered while various sources of energy enter and leave the public spotlight (1982: 27-28).

The energy utopia cycle has meant that the advantages of new energy sources tend to be overestimated while the costs are underestimated and the obstacles to exploitation often ignored. The cycle began with the replacement of wood, the earlier "fuel of civilization" by coal, America's first "black gold," during the "age of steam" in the 19th century. It was apparent in expectations that the petroleum age and its related technologies would bring forth an "era as distinct and creative as that brought by steam" and change the face of America's social and physical landscape so as to bring the amenities of urban life within reach of all (quoted in Still, 1974: 390). It continued, with the boundless prospects associated with the electricity revolution beginning in the late 19th century and extending throughout the 20th century. It was carried forward in the 1930s and 1940s when hydropower was touted as "white coal," "clean coal" and "super power" that could produce energy "in its finest, even spiritual form - electricity" (Basalla, 1982: 31). And, it has been dramatically extended in recent decades by the expectation that nuclear energy, the ultimate source of megapower, would provide the final solution to energy scarcity.

Indeed, nuclear-generated electricity continues to be pronounced as the natural energy implication of postindustrial progress.

The energy utopia cycle has meant that each new pretender to the status of energy prominence must offer not only the prospect of inexhaustible abundance but also the promise that it will usher in a new and beneficent social age. Thus, for example, recent advocacy of a solar energy future has frequently been framed around utopian visions of a Solar Age that purportedly confirms solar's indisputable superiority over other energy options. Recent literature contains guarantees that a solar energy transition will bring forth all of the following: the triumph of democracy; the accommodation to human scale; the achievement of local self-determination; the return to material simplicity and self-reliance; and the maintenance of ecological harmony. Some advocates go even further and proclaim that the Solar Age will fulfill the "ultimate moral imperative," by expressing mankind's recognition and reverence for its fixed natural endowment and by embodying a responsibility to preserve that endowment for future generations (Rifkin, 1980: 259-260).

While the claims made for a solar future are often dramatic, solar remains a minor contender for the role of reigning energy utopia. That title goes indisputably to the Electric Society, and for all the misgivings expressed over the last decade, the key to the electric society is still nuclear energy. What fossil fuels meant to the process of industrialization, electricity is now understood to mean for a postindustrial future. Indeed, it has generally be taken for granted that a postindustrial society is an electric society and that nuclear generated electric power is the ultimate high-technology path to postindustrial progress. The power of the electricity mystique has been such that it has hardly been diverted by the growing recognition of nuclear energy's cost and risks. From the vantage point of postindustrial progress, the transition to a nuclear electric society may be acknowledged as more costly than previously anticipated, but all alternatives to proceeding on course are considered even more costly. For example, Peter Navarro argues that "GNP cannot grow at a rate much higher than the rate of increase in electricity supply...utility capacity is the locomotive of the American economy" (1983: 93). Navarro further asserts that it will take at least $350 billion in capital investment within a ten-year period to keep that locomotive moving; "just to keep the lights on and the economy growing" (1983: 87). In the context of the electricity mystique such an investment is simply a necessary price of progress. Recent advertisements of the U.S. Committee on Energy Awareness (the lobbying arm of the nuclear utility industry) typify the reasoning that the Electric Society transcends economic evaluation. The American public is warned that "a healthy national economy needs a secure supply of electric energy" and that nuclear electricity is our best hope for the future. Indeed, a link is drawn between expanded production of nuclear power and conservation of domestic energy resources, leading the committee to dub nuclear "the conservation fuel." The question is thus posed: "In a competitive world market where abundant energy is a must, can America afford to fall behind in the very technology we pioneered?" In the search for the abundant energy machine, no price is too great.

Technology and the Displacement of Politics

The energy-civilization equation and the myths of megapower reflect the dominant beliefs of a technocratic culture. The most significant of these is the belief that technology is the essential instrument of social progress and that the limits of technology are generally synonymous with the limits to progress. In this context, the fulfillment of U.S. energy needs has been conceived as a technological problem requiring a technological solution. It is this conception, even more than the specific interests served by technoscience, which has shaped the politics of energy R&D and constrained evaluation of energy prospects and possibilities. The search for an abundant energy machine has thus been sustained by a self-sealing logic that presumes technology to be the pacing element of social change and that defines progress as a succession of technofixes.

In some respects, the technocratic logic that has guided the politics of energy R&D may be seen as both justifiable and prudent. In Western societies, technological advance has indeed been a measure of social progress for at least two centuries. Moreover, there is an undeniable truth in the notion that ignorance of and nonadaption to the technological requisites of modern society will only court disaster. As Langdon Winner points out, "certain technical means stand at the very basis of human survival. Failure to provide for them is to invite discomfort, suffering and even death" (1977: 102). Beyond this, it must be admitted that the challenges of modern society typically arise in the context of technological change and that meeting such challenges entails a need to directly address their technological implications. Surely the energy crisis is such a challenge; to ignore its technical dimensions or the role of technology in its outcome would be, at best, foolhardy. An R&D policy that addresses the technological character of energy choice and seeks to expand the possibilities for choice may thus be judged essential. Indeed, the basic rationale for government R&D policy cannot be reasonably separated from recognition of the society's need to meet the technological challenges of the modern era.

The reasons cited to justify the technocratic logic guiding R&D policy also signal its threat. Precisely because technology can be recognized as essential to social existence, it is possible in contemporary society for it to be regarded as a type of moral standard. Technique, through its association with survival and comfort, furnishes root meanings of good and bad, necessary and unnecessary, appropriate and inappropriate (Ellul, 1964). Indeed, we regularly judge our social options according to various attributes of the technical means for bringing them about; "the rule of technological circumstances in the modern era does in fact supplant other ways of building, maintaining, choosing, acting and enforcing, which are more commonly considered political" (Winner, 1977: 237). In this sense, technology offers not merely a means for fulfilling an end but also a framework of moral and political judgments that defines social needs and possibilities. It is distinctive of a technological civilization that such moral and political judgments tend to be shielded from open evaluation precisely because they are framed in the context of technological instrumentality. Political choice is sealed *within* technical choice and as a result insulated from external standards of value and performance. Not just any political culture will do. As Ellul has argued, technique embodies its own rationale and demands to be evaluated on its own terms.

> [I]ndependently of the objectives that man pretends to assign to any given technical means, that means always conceals in itself a finality which cannot be evaded. And if there is a competition between this intrinsic finality and an extrinsic end proposed by man, it is always the intrinsic finality which carries the day...If we make use of technique, we must accept the specificity and autonomy of its ends, and the totality of its rules...Technique does not accept the existence of rules outside itself, or of any norm. Still less will it accept any judgment upon it. As a consequence, no matter where it penetrates, what it does is permitted, lawful, justified (1964: 141, 142).

The technological preemption of moral and political choice typifies the politics of energy R&D during the past four decades. Again, no case is more instructive than the promotion and development of nuclear power.

That nuclear power is essential to social progress and in this sense good for society was decided before its practicality as an energy source was determined and before the technical means to control it existed. Nuclear power did not surface to fill an immediate social need. No credible economic or energy reasons compelled the use of nuclear power. This is not to suggest that economic and other factors did not affect or even stimulate technological innovations in nuclear power. Rather, the point is that technology generally and nuclear technology specifically has, through its political standing in Western civilization, acted upon economic life as much or more than it has been acted upon. As Winner notes, the recognition of a moral and political character in technology does not deny the existence of economic or other forces. "It merely suggests that such factors may not be the overwhelmingly decisive ones. There is a sphere of vital concern that one misses if every question is reduced to categories of economics" (1977: 236).

In the case of nuclear power, economic justification was irrelevant to its initial promotion. Nuclear power was not the invention of enterprise; there was no market demand for it and there was no economic supply of it available prior to its institutionalization. As an energy source, nuclear power was not technologically available and could not be demonstrated when the U.S. undertook its promotion. An institutional framework for promoting nuclear power was initiated with the Manhattan Project, which brought together three of the most powerful institutions in American society - the military, large industry and technoscience - to create nuclear bombs. This apparatus was then authorized in 1946 by the U.S. Congress to deliver the peaceful atom, eleven years before the idea could be technologically expressed and seventeen years before the successful demonstration of the first turnkey plant. In effect, this society set about to discover and affirm the advantages of nuclear power and to discount its costs without any knowledge of its economic practicality and before the technical means to deliver it existed. The social desirability and technical viability of nuclear power were evaluated *in the context of its active promotion* and not the other way around.

The promotion of nuclear power can only be explained in the context of technocratic norms and aspirations. Nuclear power exemplifies the modern vision of human progress. Nuclear power was and to a significant extent continues to be seen as the natural and inevitable fulfillment of the technological promise; by dreams of reason, humanity discovers the secret of energy by which all technological invention is possible and upon which human progress fundamentally depends. Through nuclear power, we would have the endless energy to create. While the computer, the laser and other modern inventions impart a sense of social awe, they are derivative of the energy secret. They do not offer limitless energy but depend upon it. In this way, the promise of nuclear power is inseparable from the promise of technological civilization itself. As one nuclear advocate has put it, nuclear power is the essential foundation for carrying forward the advances of the past two hundred years; "nuclear is the only nonfossil energy source that will be available to us in sufficient amounts to support our current civilization and to fuel progress for the foreseeable future" (Agnew, 1983: 1). Cost is inconsequential when civilization is at stake. To deny ourselves access to the energy secret is to forego progress. The enthusiasm for nuclear power, the institutional resources dedicated to its development, and the resistance to any deviation from the nuclear commitment can only be understood in the context of a ruling ideology of technological optimization.

So long as the aspirations of technological civilization define nuclear's promise, external indicators of performance are irrelevant. In this sense, nuclear power has been demanded to meet the requirements of an established technological path and, however precarious its performance may be by other standards, it is valued as a response to the systemic needs of that path. That is why the active promotion of nuclear power in the 1950s, despite its lack of availability, was deemed not only pragmatic but inevitable. It was anticipated and affirmed because it could be recognized as a logical extension of current technical and institutional arrangements for the production and use of energy. It was recognized as a practical option during the last 40 years in the same sense that, as Ellul suggests, the interconnection of electrical networks was:

> Electrical networks may remain for some time independent of one another. But this situation cannot last when it is found that independence gives rise to general costs of no inconsiderable magnitude, difficulties in arranging the courses of the lines, and even practical difficulties in electrical technique. The interconnection of electrical networks is demanded by all technical men (1964: 237).

In the context of such recognized system demands, matters of social purpose become recast as matters of efficient technical practice. As Winner has pointed out, "the technological society tends to arrange all situations of choice, judgment and decisions in such a way that only instrumental concerns have any true impact. In these situations questions of 'how' tend to overpower and retailor questions of 'why' so that the two matters become, for all practical purposes, indistinguishable" (1977: 232-233).

For four decades the central question of energy choice has been defined as instrumental rather than purposive: essentially, how to perpetuate the "rapid expansion of centralized high technologies to increase supplies of energy, especially in the form of electricity" (Lovins, 1977: 25). The basic issues have been framed in terms of the technological requirements of the existing energy path. Two types of technical requirements have been dominant. First, the technology must be compatible with the scale and centralization of existing energy production and distribution; it must meet the condition of connectivity with prevailing technical and institutional arrangements. Second, it must be suited to the delivery of energy in a form currently forecast as essential to continued progress along the existing technological path; it must meet an Electric Society's definition of technical efficiency. Within the technological paradigm, it is virtually undeniable that nuclear power, and possibly *only* nuclear power, may meet both of these requirements. Once this judgment is made, the only remaining question is, how can we make the nuclear energy machine work?

To a significant extent, energy R&D policy has been framed around this question. The commitment to nuclear power has been treated as practically irreversible. The growing catalogue of nuclear's liabilities has grudgingly been admitted, but it has not led to a fundamental alteration of our technological path. Instead, we have engaged in a process which Winner calls "reverse adaptation" whereby "a series of transformations engendered by the introduction of technological means eventually leads to transformation of ends"; "ends are adapted to suit the means available" (1977: 238). Thus, for example, the declining economic fortunes of the nuclear industry are not to be regarded as an indictment of nuclear economics but, rather, a compelling justification for increased capital investment. Similarly, issues of nuclear proliferation, nuclear waste disposal, and nuclear safety are acknowledged to be serious, but we must recognize that all progress has unanticipated consequences. These consequences must now be treated as simply a new generation of technological challenges requiring an even greater investment of resources in finding technological solutions. In the spirit of such contemporary challenges, breeder reactor technology is held out as a *solution,* incredibly enough, to the threat of nuclear proliferation. Similarly, technical rationality dictates that the waste disposal problem is to be resolved, for example, by transporting the Shippingport reactor core by barge down the

Atlantic coast, through the Panama Canal, and up the Pacific coast to the Hanford reservation in Washington. And the sensible approach to safety issues is to cap the accident liability of the utility industry until publicly supported nuclear R&D - which, logically, must be increased - yields acceptable answers; under such a blaming-the-victim strategy, the public must pay for addressing the threat to public safety posed by the technological choices of the utility industry. In sum, we may conclude that the Electric Society is more costly and dangerous than previously imagined, but we dare not risk an alternative. To reason otherwise would be to raise suspicion about the stability of our technological civilization. In fact, to reason in any other way would be to admit technological defeat.

The search for the nuclear energy machine demonstrates the compulsive quality of technological choice. As Ellul has said, technique accepts no permanent limits: it "is autonomous and recognizes as barriers only the temporary limits of its action"; "it is driven onward by itself, by its character of self-augmentation" (1964: 142). Once launched upon a technological path, obstacles can only be evaluated as new challenges to be overcome by enhancing our commitment; technique provides its own moral and political justification. Recognizing this should draw our attention to Amory Lovins' warning that we stand at a crossroads of energy choice and that without decisive action our options will slip away. As he has argued the differences between the hard path we are on and the soft path that is available "may seem abstract, but they are sharp, imminent, and practical" (1977: 59). Those differences are, most importantly, social, political and ethical. "Crucially at stake," according to Winner, "are choices that concern the essential character of energy regimes: how large they will be, what energy sources they will employ, how they will be designed, how and by whom they will be run, and how they will affect the structure and texture of human life" (1982: 275). These matters cannot be evaluated if technical imperatives are accepted as surrogates for moral and political judgment. Unfortunately, the primary and most disconcerting legacy of four decades of energy R&D policy is the acceptance of such imperatives.

R&D Policy and Energy Regimes

While not usually analyzed as such, energy R&D policy has been an instrument of the current American energy regime. That regime is dominated by economic and political institutions which are militaristic, highly centralized, technocratic, and dedicated to the preservation of a culture of energy abundance. The objective of this regime has been an energy future governed by large-scale technical systems. This is evidenced most strikingly by the overwhelming and virtually uncontested R&D support given to nuclear power. To a significant extent, energy policy in general and R&D policy in particular is now driven by the inertia of the nuclear commitment. This commitment has been reinforced by a prevailing mythology that equates energy use with civilization and defines social progress as the achievement of the Electric Society.

Under the current regime, opportunities for conservation and the use of renewable energy will be undervalued. R&D policy will continue to search for the abundant nuclear energy machine. Only those technical solutions consonant with existing energy institutions will be judged as feasible. The inversion of technical means and social ends will persist, not because it is desirable, just or freely chosen, but because it is deemed necessary.

REFERENCES

Agnew, Harold B. 1983. "Civilian Uses of Nuclear Power: Status and Future." *Proceedings of the Symposium on Energy: Challenges and Opportunities for the Middle Atlantic States* 2. Baltimore, MD: Johns Hopkins University, Energy Research Institute.

Basalla, George. 1982. "Some Persistent Energy Myths." In George H. Daniels and Mark H. Rose (eds.), *Energy and Transport: Historical Perspectives on Policy Issues.* Beverly Hills, CA: Sage Publications.

_______________. 1980. "Energy and Civilization." In Chauncey Starr and Philip C. Ritterbush (eds.), *Science Technology and the Human Prospect.* New York, NY: Pergamon Press.

Bush, Vannevar. 1945. *Science, The Endless Frontier.* Washington, DC: U.S. Government Printing Office.

California Energy Commission. 1981. *Energy Tomorrow: Challenges and Opportunities for California.* Sacramento, CA: California Energy Commission.

Commoner, Barry. 1979. *The Politics of Energy.* New York, NY: Alfred A. Knopf.

Congressional Budget Office. 1984. U.S. Congress. *Research and Development Funding in the Proposed Fiscal Year 1985 Budget: Special Study.*

Ellul, Jacques. 1964. *The Technological Society.* New York, NY: Vintage Books, Random House.

Guerra, Joe. 1981. "Nukes Face Bottom Line." *Working Papers,* 8/6 (November/December): 20-21.

Hewlett, R. G. and B. J. Dierenfield. 1983. *The Federal Role and Activities in Energy Research and Development 1946-1980: An Historical Summary.* Oak Ridge, TN: Oak Ridge National Laboratory.

Hill, Christopher. 1985. "Rethinking Our Approach to Science and Technology Policy." *Technology Review* (April): 11-15.

Lakoff, Sanford A. 1969. "Scientific Society: Notes Toward a Paradigm." In Paul Piccard (ed.), *Science and Policy Issues.* Itasca, IL: Peacock.

_______________. 1964. "The Third Culture: Science in Social Thought." In Sanford A. Lakoff (ed.), *Knowledge and Power.* New York, NY: Free Press.

Lambright, W. Henry. 1976. *Governing Science and Technology.* New York, NY: Oxford University Press.

Lovins, Amory B. 1977. *Soft Energy Paths: Towards a Durable Peace.* New York, NY: Harper and Row.

Morell, David. 1981. "Energy Conservation and Public Policy: If It's Such a Good Idea, Why Don't We Do More of It." *Journal of Social Issues* 37/2: 8-29.

Navarro, Peter. 1982. "Our Stake in the Electric Utility's Dilemma." *Harvard Business Review* (May-June): 87-97.

Nelson, Richard R. 1977. *The Moon and the Ghetto.* New York, NY: W.W. Norton and Company.

Odum, Howard T. and Elizabeth Odum. 1976. *Energy as a Basis for Man and Nature.* New York, NY: McGraw Hill.

Price, Don K. 1965. *The Scientific Estate.* Cambridge, MA: Harvard University Press.

Rhodes, Steven L. 1983. "Implementing Federal Solar Policy: Learning from the Nuclear Experience." *Journal of Energy Law and Policy* 3/2: 189-241.

Rifkin, Jeremy. 1980. *Entropy: A New World View.* New York, NY: Viking Press.

Sapolsky, Harvey. 1975. "Science Policy." In Fred I. Greenstein and Nelson W. Polsby (eds.), *Handbook of Political Science: Policies and Policy Making* 6. Reading, MA: Addison-Wesley Publishing Company.

Shapley, Willis H., Albert H. Teich and Jill P. Weinberg. 1983. *Research and Development, AAAS Report VII: Federal Budget--FY 1983, Impact and Challenges.* Washington, DC: American Association for the Advancement of Science.

Still, Bayard. 1974. *Urban America: A History with Documents.* Boston, MA: Little, Brown and Company.

Ward, Barbara. 1977. "Foreword." In Amory Lovins, *Soft Energy Paths: Towards a Durable Peace.* New York, NY: Harper and Row.

Winner, Langdon. 1982. "Energy Regimes and the Ideology of Efficiency." In George H. Daniels and Mark H. Rose (eds.), *Energy Transport: Historical Perspectives on Policy Issues.* Beverly Hills, CA: Sage Publications.

_______________. 1977. *Autonomous Technology: Technics Out-of-Control as a Theme in Political Thought.* Cambridge, MA: MIT Press.

Yergin, Daniel and Martin Hillenbrand (eds.). 1983. *Global Insecurity: A Strategy for Energy and Economic Renewal.* New York, NY: Penguin Books.

Energy Policy in the Interim: Waiting for the Next Shoe to Drop

Grant P. Thompson

The dry spell for energy policy continues with little sign of return to the days when oil prices rose monthly and energy experts found government grants and invitations to TV talk shows equally easy to arrange. Whether measured in dollars spent for civilian research by the U.S. Department of Energy or in attention paid to the subject by the media and environmental and consumer organizations, the energy crisis has faded.

To a very great extent, the new public indifference is a blessing. Those who made a profession of forecasting energy prices have seen their best guesses undercut by a glutted world market. Too much public attention could only be embarrassing if memories were even a little better than the predictions had been. On the plus side, government policy has finally embraced the recommendation that formed the central message of virtually every energy study made during the 1970s: we now accept that it is a better approach to use the market to allocate energy among competing users rather than depending on government to decide who should get oil and natural gas (see Landsberg et al., 1979; Schurr et al., 1979; Schelling, 1979; and Glasner, 1985).

Shouldn't those of us who have been active in advising government simply declare victory and move on to solving some other social problem such as water allocation in the west or peace in our time? In this article, I will argue that, despite the very great strides the United States has made toward a more rational energy policy, there is still more work to be done before we remove the subject from policy debate. Perhaps more provocatively, I will argue that the almost aggressive indifference of the federal government towards energy policy is sadly shortsighted in a number of important ways.

Energy policy would be of little interest if we were sure there would be no future shortages of energy. Yet the trend lines suggest that we may not be so lucky. The vast majority of exploration and production has taken place in the Western developed nations; the greatest known reserves still lie in the Middle East. Nor is it realistic to believe that we will be able to make quick substitutions of alternative fuels if oil prices rise. And no one is confident that electricity can penetrate the transportation market in the near future. Thus we will have to continue to be concerned about fossil energy sources for the foreseeable future. The question is whether we should leave the worrying and planning solely to market forces or whether there is a role for government to play.

Triumphs of a Decade of Policy

Before we examine the work that remains to be done, it is worth celebrating the progress of a decade or more of public policy in the energy arena. A look at our ability to learn from our mistakes may blunt the criticism that academic and professional research has no effect on the real world. There are three improvements in public policy that have occurred in the energy field in the last ten years: removal of price controls and allocation systems; separation of energy pricing policies and welfare issues; and a more sophisticated understanding of the relationship between energy use and productivity.

Price Controls and Allocation Rules

Recall the situation in the early 1970s. In an effort to control the inflation that accompanied the wind-down of the Vietnam War, President Nixon imposed wage and price controls on most items in the economy. At around the same time, a tightness in the world oil market developed, making it possible for the first time for the Middle Eastern oil producing nations to cut production and raise prices. Although it was generally agreed in this country that the wage and price controls had not worked to slow inflation, nonetheless, when they were lifted for most other items, the President left them on petroleum. So while energy prices rose in other parts of the world - thus rationing oil to those who were willing to pay for it by spending more money - the U.S. government developed an allocation system for industrial, commercial and residential users. The result was that we paid for oil by spending time in gas lines or by lobbying for special preferences. Being willing to pay more earned no preference in the fight for limited supplies of oil. Perhaps the most surprising fact about this unhappy state of affairs was the number of people who defended the misery caused by government regulations and sought to correct it by more intricate rules.

The shift in perception that allowed oil decontrol legislation to be enacted probably grew out of a combination of several factors. The oil companies worked hard through advertising and lobbying to press this point of view, although their obvious self-interest almost certainly blunted the effectiveness of their educational efforts. The demonstrated ineffectiveness of the federal and state allocation bureaucracy - who were charged by Congress to do the impossible and tried as best they could to comply - made many aware that allocations and price controls did nothing to lower world prices and had done very little to reduce hardship within the general population. Economists and academics were effective at pressing their message concerning the use of the market to allocate scarce resources. Finally, the political courage of President Carter must not be discounted, for it was his administration, not the Reagan administration as many believe, that decontrolled oil prices and began the process that has led to decontrol of natural gas wellhead prices as well.

The first triumph of policy during the decade was, then, the growing realization that no set of regulations could be as fine-grained as the market in making the millions of allocation decisions that our large economy demands. The legislation passed in 1978 that set in motion the eventual decontrol of oil prices was a wise admission that markets are better at allocating items of commerce than government regulations. The U.S. thus rejoined the world economy for oil and the conservation reaction induced within this country played an important role in easing the oil market for all nations.

Energy Prices and Welfare Policy

Also during the decade, policy makers in this country accepted the notion that energy policy and welfare policy are different issues. Easing the misery of the poor should be carried out directly as a function of a civilized society rather than hidden within an energy policy that tries to keep consumer prices low for everyone, whether rich or poor. Trying to combine welfare and energy policies results in a distorted system that fails to help all who genuinely deserve assistance while simultaneously rewarding many who do not need help.

Representatives of the poor feared that if energy prices were freed from control, welfare policies would not be made more generous; events have shown they were right. In fact, one can argue that the decoupling of these two issues was simply part of the growing social perceptions that the poor deserve less support overall and that many who claim to be poor are simply lazy. In any case, for whatever reason, the poor have both higher energy prices and lower welfare payments. It is easy to argue from principle if one is rich; but when lower energy prices appear to be the only popular way to deliver some help to hard-pressed families, we should not be surprised when decontrol is resisted. But this may be an example of where the U.S. has reached the right policy result for the wrong reason. I will return to this point later in the article.

Energy Consumption and Economic Productivity

The third great breakthrough in policy made in the last decade was that public policy makers came to accept that higher prices change the behavior of both suppliers and consumers of energy. The case for higher prices as a way of encouraging more exploration and drilling was made repeatedly by oil and gas suppliers. But during the early 1970s, only a few people argued that the other side of the equation was true as well: higher prices would improve the efficiency of energy use (see, for example, Energy Policy Project of the Ford Foundation, 1974). The decade reminded us without any question that energy demand is not fixed in relation to economic growth. We learned that, unlike former Secretary of Energy James Edward's claim, there were more ways to escape from the energy crisis than simply to "produce, produce, produce" *(Business Week,* 1981).

This is a lesson that we should have known already, in view of the variations experienced during earlier periods of recent history. For example, expressed as a ratio between units of energy consumed for every unit of Gross National Product created, we have seen shifts from 0.63 (during 1950-55) to a high of 1.45 (from 1965-70) (Franssen, 1979). But the point had to be argued repeatedly during the early years of the energy crisis. We now understand that the economy can grow at a satisfactory pace using less energy per additional unit of output than had traditionally been believed. For example, a 1983 Office of Technology Assessment (OTA) study observed that "between 1972 and 1981, American industrial energy use declined by over 2 Quads, and energy efficiency improved by almost 18 percent per unit of production" (OTA, 1983: 4). An earlier OTA analysis of the residential sector noted that households did not have to forego amenities in order for the rate of energy use in this sector to be slowed:

> Americans as a group were increasing their use of energy in the home at an average rate of 4.7 percent per year; in the 1970s,

> the annual growth rate has averaged 2.6 percent. Moreover, the remaining growth is attributable primarily to a growth in the number of households; the amount of energy used in each household has remained almost constant between 1970 and 1977 (OTA, 1979: 4).

Similar findings occurred in an examination of energy consumption in the transportation sector (OTA, 1982). In sum, letting prices rise in response to increased demands encourages at least some consumers to find alternatives, while it simultaneously tempts at least some suppliers to find new, higher priced, sources of supply.

These three new-found realizations now seem so self-evident that it is hard to remember how long it took for Congress and the public to accept them. But they represent a real triumph of careful, repeated examination of the issues by persons trained in public policy analysis and a sustained process of communication to the public.

Unfinished Business

The successes of policy implementation noted above do not exhaust the work that must be done. In the remainder of this article, I outline five additional policy issues that must be attended to before we can call our work complete. These issues are (1) to fulfill the agreement we made with the poor on energy prices and welfare; (2) to begin the reform of electricity pricing; (3) to support energy conservation initiatives that extend beyond normal market horizons; (4) to distinguish between research and commercialization, then get on with research; and (5) to make a sensible peace between environmental regulations and energy development.

A Parenthetical Digression on Natural Gas

In writing such an article a year ago, I surely would have included a section on the need for natural gas price reform. A year and more before, the mispricing of natural gas loomed as one of the major energy policy issues left unresolved by the 1970s. But circumstances have changed enough so that the theoretical disparity between market-determined prices and actual prices has become less important. Aficionados of the complicated may recall the two dozen plus categories of natural gas created by the Natural Gas Policy Act of 1978; some may still remember that the gas we are buying today is a mixture of "old" gas (still under price control) and "new" gas (almost entirely released from price regulation as of January 1, 1985). There was great concern that there would be a price "fly-up" when the price of new gas was decontrolled. Some people suggested that price caps should be imposed or that all new gas be resubjected to price controls (see House Committee on Energy and Commerce, 1982). Only the memory of the bloody battles that went into the 1978 Act prevented Congress from reopening the issue. As it happened, a variety of circumstances, including a warm winter and slackening of industrial demand combined to take the pressure off natural gas supplies, so that prices did not move dramatically upon decontrol.

Although there is still a powerful economic argument for removing the remaining price controls from "old" natural gas, the practical argument is less compelling. The "old" gas will be depleted before the end of the decade and its distorting contribution will be relatively unimportant well before the political will could be mustered to remove regulations. It is a case where the passage of time will serve to remove both the problem and the pressure to solve it.

There is, to be sure, an important issue remaining: whether natural gas pipeline companies should be taken out of the business of buying and selling gas and turned into either common carriers or contract carriers. Pipeline companies, which now purchase the natural gas at the wellhead and sell it to distribution companies at the city gate, may have somewhat less of a short-term incentive than the ultimate consumers do to hold down prices. If pipeline companies became simply transporters of gas for pay - something like a trucking company - some would argue that distribution utilities and large industrial customers would go to the fields themselves and bargain more vigorously with gas producers for lower prices or better conditions.

This argument is more persuasive for many observers when it is noted that a number of pipeline companies are also natural gas producers. This arrangement provides opportunities, it is charged, for the dual-function companies to shift profits between the regulated pipeline and the unregulated producer subsidiaries. Many in the industry dispute this view (see, e.g., Interstate Natural Gas Association of America, 1984). However, there is enough analysis on both sides of the issue to warrant consideration of the question (see Energy Information Administration (EIA), 1984; and Broadman et al., 1983).

Although the issue of the status of pipeline companies is important, it will, in my judgment, play itself out in the normal course of political discussion. The issues are complex and not very susceptible to public debate and the experts themselves disagree on which solution would permit the natural gas market to approach the ideal of a free market most closely. Thus, important as the natural gas market once was as a subject of policy, and important as the status of pipelines may become, it is not among the five issues I now propose to discuss.

Energy and Welfare

I have noted above that much of the fight over decontrol of oil and natural gas was conducted within the rhetorical cloak of protecting the poor and those living on fixed incomes from rapid price increases. Although one might have suspected from time to time that those who wrapped themselves in that mantle really had their own self-interest at heart, there can be no doubt that many people were honestly concerned how families that spend large portions of their income on necessities would cope with increasing heating bills. Stories of welfare recipients forced to choose between heat and food were as compelling as they deserve to be.

In 1980, it was calculated that the poor as a group were forced to bear some $3 to $7 billion decrease in their annual real income as a result of energy price rises that were uncompensated by increased welfare payments (Committee for Economic Development and The Conservation Foundation, 1982). Since 1980, energy prices have moderated, but welfare payments have been cut dramatically, so that the figure is probably higher yet. Because the poor live disproportionately in older, less well insulated homes and apartments and do not have the capital necessary to purchase more efficient heating, cooling, or refrigerating equipment, their ability, as a group, to defend themselves against higher prices is limited.

Yet the advantages of price decontrol finally won out over the opposition of representatives of the poor. Those of us who argued for letting the market work noted that welfare issues and price issues should be treated separately and there was at least an implied promise that those benefiting from price decontrol would then help make the case for welfare payments that took into account higher energy prices.

In an article about energy policy, I will not digress at length at how such welfare ought to be delivered to help those most in need. But it would be disingenuous of those among us who argued that the poor should support price decontrol to walk away from the issue now without at least considering whether we have some obligation to alleviate the additional suffering that higher energy prices have caused during the last decade. Until that issue is addressed in a responsible manner, it is hard to feel that energy policy is closed for the decade.

Electricity Pricing

The single most important facet of energy policy that we have left untouched is reform of how we price electricity to the ultimate consumers. Coal, oil, much of natural gas, wood, and solar energy all are priced by market forces. But electricity continues to be regulated and priced in a way that almost certainly guarantees that we will be locked into a cycle of building more power plants than we really want or need and then paying for them.

Recall that economic theory suggests that commodities and services should be priced at their marginal cost, that is at the cost of producing one additional unit of the item. Consumers then have an accurate idea of the resources that society must devote to making more of the item and can decide whether, at that price, they want more or less of that item rather than more or less of something else. It is this way of establishing prices that, ideally, leads to an efficient allocation of resources throughout an economy.

In contrast to this system, electricity is priced under public regulation in a way that calculates the sum of the total historical cost of the plant used to produce the power (less depreciation, of course) plus the variable costs of fuel and operating costs and simply divides the total by the number of kilowatt-hours sold to set the price of the power. Not surprisingly, there is only a random relationship between this average price of power and the marginal price of power that ought to be the measure of cost. Because most power is generated from plants that were constructed a number of years ago when real costs and interest costs were both lower, the result of this method of pricing electricity is that it is sold too cheaply. As with any commodity sold below its marginal cost, it appears to the consumer to be a better deal than it really is (in terms of the resources devoted to its production) and consequently consumption rises.

A comparison of the price and the consumption of electricity with other energy sources during the last decade illustrates this point. Although the price of all energy forms rose during the decade, in real terms electricity prices rose far less than oil or natural gas. In constant dollars, between 1970 and 1980 gasoline prices increased 71 percent; residential heating oil prices increased by 171 percent; wellhead natural gas prices increased by 368 percent. However, residential electricity rates increased by only 33 percent; and industrial electricity rates have increased by 60 percent (EIA, 1983: tables 43, 51 and 72). This means that, compared to other fuels, electricity is a better bargain than it was 15 years ago.

Consumption figures suggest that consumers are responding as we might expect to this change in relative prices. Although electricity consumption has grown at a slower pace than in the 1960s, it has still increased at a faster rate than oil or natural gas. A recent EIA study indicates that electricity's share of the residential fuel mix tripled between 1960 and 1982, and doubled in the commercial and industrial sectors (EIA, 1984: 18, 24 and 34). A number of observers believe that, although total energy usage in the year 2000 will be less than conventional estimates, the electricity share of that consumption will be larger than now. Indeed, many project that even with a slowing down in the growth rate of energy demand, electricity's share of the fuel mix will rise (Loftness, 1984).

Thus, the underpricing of electricity from an economic point of view appears to be encouraging consumers to use more of the commodity than they otherwise would be inclined to do. In making those additional electricity purchases, they are forgoing purchase of more efficient lighting, higher efficiency air conditioners and refrigerators, better motors, and improved use of daylighting (to name only a few of the thousands of substitutes that exist for present day electricity use).

The problem with changing the method of calculating the price of electricity is that society has devised a closed system of rates. Electric utilities are monopolies that receive a limited rate of return in exchange for an exclusive franchise.

While marginal cost pricing would produce more revenue than a traditional system based on historical costs, there is no way to permit the company to collect that money from ratepayers and then prevent the company from keeping it. Regulators are faced with two choices: they can leave electric rates lower than economic theory would dictate and face the long-term prospect of overbuilding generation facilities to meet the demand that such underpricing stimulates; or, alternatively, they can raise the rates and try to figure out some way of returning that money to consumers in a way that will not simply encourage the same over-consumption. It is not surprising that commissions that have faced this dilemma have chosen to avoid the intellectual and political difficulties of raising rates: "At times, public utility regulators must feel that they have been asked to untie a Gordian knot" (Gormley, 1983: 4).

In principle, the solution is quite straightforward. The object is to devise a system that does not permit a regulated monopoly to prosper yet does not provide electricity consumers with a service priced below its replacement cost. We could simply set rates according to economic theory while at the same time calculating the amount of money that the utility would have been permitted to collect under traditional historical average cost rates. The difference between the figures would simply be taxed away by the state. (This fund would be an attractive source of revenues to support increased welfare payments, of course.)

Yet we can imagine how politically unpopular such a system would be with all customer classes. The fact that it would require legislative changes to carry out gives that political unpopularity real teeth. The average middle income consumer would not see any discernible advantage from the higher rates, while the industrial consumer would see costs rise in this country in a way that satisfied economists but not salesmen. It is the nature of economics that it delivers its bad news in large lumps and its good news only over time and in small dollops.

Even though moving prices closer to marginal costs is extremely difficult as a political matter (some would say impossible), the case needs to be made strongly by policy analysts. The price of not making the case will be a massive misinvestment in electric power plants that could run into billions of dollars (e.g., compare Navarro, 1982; and Congressional Research Service, 1983).

There are at least some half-way measures that can be taken right away to help substitute for a genuine reform of electricity pricing. Since the object of letting electricity prices rise to their marginal cost is not to punish the consumer but rather to provide incentives to find less expensive ways to achieve the goals that electricity service now satisfies, we might ask the electric utility to buy more efficient equipment for the customer up to the point where the money expended on alternatives equaled the marginal cost of providing additional power. This is the line of reasoning that lies behind utility programs that pay consumers to buy more efficient refrigerators, that provide low-cost loans for insulation, or that give audits and free advice to customers on efficient electricity use. Commendable as these programs are, the total spending devoted to them at this time is pitifully small when compared with the cumulative savings that might be achieved if even a few large central station generating plants were canceled or postponed and resources devoted instead to more efficient buildings. Thus, the regulatory commissions and the companies need to be bolder in assessing the savings to customers from such investments and then devoting resources commensurate with those savings to buying efficiency for the customer; to think in terms of what Arthur Rosenfeld of Lawrence Berkeley Laboratory has called "the supply curve for conserved energy" (Meier et al., 1983).

Increased support by the utility industry for research and development on improved efficiency is another practical halfway house toward using money collected in rates to substitute for actions that consumers and suppliers might take if electricity prices were higher. In theory, of course, the electric utility industry should engage only in research and development that directly benefits the industry itself. We would ordinarily expect light bulb makers to do research on their products and boiler makers to improve theirs, while utilities worried about better construction management or less costly distribution systems. If electricity prices were at marginal cost, this simple model would be correct, since each party would find the incentive to undertake research for the rewards it brought to the sponsoring enterprise.

But that incentive is lacking to the extent that electricity is priced below its replacement cost. No inventor will receive the reward he or she deserves if electricity prices fail to prod users to buy the new product. Failing that incentive to invent and produce more efficient equipment, we might logically look to the utility industry to provide at least some of the missing R&D out of the funds it collects from its ratepayers. Filling in the gaps that lower rates create provides the justification for end-use R&D by an institution such as the Electric Power Research Institute (EPRI).

In this connection, it is heartening to see that the Board of Directors of EPRI in their recent ten-year effectiveness review called on the organization to devote a larger share of its resources to end-use R&D (EPRI, 1984). It remains to be seen whether the industry and public regulators will provide the increased funding that such a program deserves.

Even at their most aggressive, programs such as conservation investments on the customers' side of the meter and R&D conducted on behalf of the makers of electricity-consuming devices are half-way, stop gap measures that tap into only a small part of the potential for electricity conservation. Although it currently seems impossible that we could someday reform the pricing of electricity in the way we did for oil and natural gas (or airlines or telephones), it remains the single most important item of unfinished business on the energy public policy agenda.

Energy Conservation Programs

Although pricing reform has been the single most effective energy conservation program this country has undertaken in the past decade, it is simply incorrect to assert, as the Reagan administration does, that prices alone will call forth all the economically justifiable conservation. The exceptions are well-known and large:

o Rental housing and commercial office space, where the incentives to conserve are split between the landlord and the tenant.

o New owner-occupied housing, where the payback period demanded by the buyer for cost-effective conservation investments is far shorter than the lifetime of the housing stock.

o Major appliances, including heating and refrigeration, purchased by a builder and installed in a home prior to sale.

o Virtually all the equipment installed in speculative office buildings.

o Any energy-consuming device where the costs of future operation are difficult or impossible for the buyer to calculate in advance.

In addition, there is the large category of items, of which the automobile is the best example, where the lead times for production of improved models is sufficiently long and costly that market forces operating in the short run may deliver messages that are contrary to the long term interests of the country as a whole. For many reasons, most people would agree that if we could import less oil it would be better for the nation's economy. Yet fluctuating energy prices may blur this longer term goal. Thus, there may be a strong argument that all of us together should buy more efficient cars, even if no one of us feels the pressure to do so.

In each of these cases, we can justify taking governmental actions because the market, working alone, fails to reward conserving actions or rewards it too slowly to achieve other desirable social goals. Nor should we be ashamed of such "departures" from market economics. Governments exist to do those things that individuals acting alone cannot or will not do. The present political climate may not favor a very wide definition of what the market cannot do, but the interests of the nation as a whole should be marshalled to support conservation programs that include efficiency standards for new buildings, strong automobile standards, appliance standards, and minimum standards for speculative office and small commercial buildings. To stop short of this small amount of regulation would be, in my opinion, to substitute orthodoxy for common sense. A consistent theme of the 1970s was the undervaluing of the contribution that improved energy efficiency could make in solving the nation's energy problems. Yet energy conservation, together with a diversification of sources and fuel types, lies at the heart of our comparative freedom from OPEC's grip now. It would be shortsighted in the extreme if we failed to be aggressive in making further improvements. Unfortunately, the current national administration's attitude toward energy conservation is not encouraging, as reflected in its repeated and partially successful efforts to reduce conservation R&D funding.

R&D and Commercialization

One of the hallmarks of government policy during the 1970s was a large commitment to spend public money to hasten the commercialization of a variety of energy technologies. Spending on everything from nuclear fusion to solar cells increased dramatically. At the same time, governmental commitments to basic energy R&D rose to new levels. The most dramatic action Congress took was the creation of the Synthetic Fuels Corporation, charged with creating a full scale industry. To underline the fact that the corporation was not an R&D program, the Energy Security Act expressly forbade the corporation from sponsoring research.

The Reagan administration declared war on both the Department of Energy and the Synthetic Fuels Corporation early in its first term. The first Reagan energy secretary made little effort to conceal his dislike of the department he headed or of his intention to define his success by his ability to abolish it quickly. The Synthetic Fuels Corporation was slated for abolition, and both the pure R&D programs and the more action-forcing programs of the department were slashed in the initial Reagan budget and every one since then. Although proponents of each energy form (solar, conservation, fossil, or nuclear) protested that their favorite had been singled out for particularly harsh treatment, in fact no part of DOE's program has been immune from reduction and redirection even though conservation and renewable energy programs were clearly especially targeted. (Only the weapons divisions continue to receive additional resources.)

In this wholesale slashing of budgets, there has been a general confusion about the difference between the proper role of the government and the role of industry in conducting R&D. Early in the first term, the department developed the rubric that government should support only "long term, high risk, high payoff" energy research. Yet the department has repeatedly proposed budgets to Congress that contain relatively little research and development. This simply invites a political process of R&D planning to take place on Capitol Hill as various committees restore favorite projects and programs.

To be sure, the line between R&D and commercialization is not always clear and it differs somewhat from one technology to another. For example, I believe that the administration is correct in its dislike of the Synthetic Fuels Corporation. That entity, born in a period of zeal to buy our way out of our dependence on OPEC by throwing money at a set of old-fashioned technologies, has proven to be the wrong way to buy new energy technologies. The last thing this country needs is an industry that can provide fuel at between $60 and $90 per barrel when we can buy it in the $25 range on the world market.

To be sure, synthetic fuel is an excellent concept that needs further research. The United States has vast coal reserves and our ability to generate biomass is equally enormous. Yet, so far, we have only succeeded in using those resources to burn for heat or to convert into electricity. We should and must look for other ways to convert the resources into useful, clean energy forms. But at no time in the foreseeable future is it likely that we will want to create at governmental expense an enormous white elephant of an industry that is capable of producing high-priced oil and gas in a world full of less expensive alternatives.

We should close down the corporation and use a part of the funds to finance cooperative government-industry research on new ways of using coal, biomass, and other plentiful energy sources to provide liquid fuel for our transportation system. This research and development program should be based in the laboratory; government funding should end at the point that a process looks promising enough for someone else to risk money on a pilot scale plant.

The savings to the government, even assuming that some new R&D programs were started to replace the Synthetic Fuels Corporation's activities, would be large. More important, the future savings to the country of having a variety of genuinely new ideas ready to be scaled up if necessary would be even greater.

The reluctance shown in the U.S. toward government-sponsored research and development is curiously ideological for a nation that prides itself on its practicality. It is even more surprising when the success stories of collaborative R&D between government and private industry are so well-known, even in the energy area. Nuclear generated electricity owes its existence to government sponsorship.

Turbines in generating plants have benefited from similar research conducted for jet engines. Advances in our understanding of how residential and commercial buildings work has been conducted by both national laboratories and through cooperative research with the building and engineering professions. Research for high efficiency ballasts for fluorescent lights was conceived and funded by federal laboratory personnel; development took place in the private sector. There is no reason that major oil and chemical companies, for example, cannot be enlisted in the business of finding alternatives to oil-derived gasoline and rewarded for their inventiveness through government assistance.

Environmental Regulations and Energy

Environmental regulations are regularly blamed for interfering with energy exploration, production, and waste disposal. A theme of the confrontational era of Secretary Watt and Administrator Burford was that the selfishness of the elite environmentalists was denying the nation the opportunity to use its own resources to free itself from OPEC. Although in more muted tones, that same theme appears still in discussions of environmental regulations.

Yet there is compelling evidence that Americans place an extremely high value on a clean environment. In poll after poll, respondents state they would be willing to sacrifice money and convenience to ensure high quality air and water (see The Conservation Foundation, 1982: 424-29; and 1984: 27-29). At the same time, the National Parks and the wilderness system are both revered as unique expressions of deeply held Ar rican values.

Attacks on environmentalists and environmental regulations obscure the arguments that developers should be making, while wasting time and resources that could better be spent on more productive activities. Each side needs to recognize in their discussions about a particular piece of real estate or a particular effluent that questions of science and values are involved. By their nature, values can differ and it is a primary function of the political process to settle those differences. Science only rarely can answer environmental questions with the kind of certainty that would satisfy a lawyer, so that even scientific questions often must return to a discussion of values. But in this case, it is the value that one places on certainty, on risk under varying circumstances, and so on. Railing against the evil energy companies or against the know-nothing environmentalists does nothing but raise the blood pressure and perhaps one's standing among those who already agree with you.

Instead, I suggest that environmental and business groups should both spend more time trying to understand the substance and the merits of positions taken by the other side and learn to recognize (if not to accept) the values that drive those opinions. The successful experiences that a growing number of individuals have had working with mediators and facilitators suggests that government, industry, and public interest groups should be bolder in trying to meet for tough but fair exchange of views and to explore the possibility that truly better solutions may be found.

Unfulfilled Promises

The decade of the 1970s was not an easy one for the U.S. and the two energy price shocks certainly contributed to its discomfort both directly, by imposing an involuntary "tax" paid to foreign oil producers that acted as a drag on economic growth, and indirectly, by highlighting the fact that even a rich and powerful nation cannot insulate itself from the world's economy. But to the extent that the decade focused attention on some underlying truths about the world energy situation - that oil resources are finite and unevenly distributed, that energy is a commodity that responds to market pressures in both production and consumption, and that government policy has a role to play where the market works poorly or too slowly - the U.S. arrived in 1980 with a far better set of energy policies than were in place in 1970.

The question, half way through the decade of the 1980s, is whether we remember those lessons and whether we are using the period of calm in the world energy situation to create more flexible and robust energy systems. It seems as if we are no longer paying attention. The country has virtually walked away from any responsibility for energy policy at the federal level, answering every complicated problem by muttering the mantra of "free market." The glamor and, more importantly, the research support from either the federal government or private foundations, has dried up so that only a handful of the cadre of bright, committed students of energy policy who were trained so expensively during the 1970s have been able to continue their research efforts. And the mindset of the society is that energy is last year's problem: relax the automobile standards; let the buyer beware when buying a new house or refrigerator; and forget trying to handle the politically treacherous business of pricing electricity to reflect its marginal cost.

These challenges are no less difficult than the ones we dealt with before. Yet somehow the will must be found to undertake them lest we face the next energy crisis with a decade of preparation time lost forever. Moving beyond ideology on all sides will play a major role in seeing where the market is best and where government policy can help.

REFERENCES

Broadman, Harry G., W. David Montgomery and Mary Beth Zimmerman. 1983. *Natural Gas Markets After Deregulation: Methods of Analysis and Research Needs*. Baltimore, MD: The Johns Hopkins University Press for Resources for the Future.

Business Week. 1981. "Energy Conservation: Spanning a Billion-Dollar Business." April 6: 69.

Committee for Economic Development and The Conservation Foundation. 1982. *Energy Prices and Public Policy*. New York, NY: Committee for Economic Development.

Congressional Research Service. 1983. U.S. Congress. *A Perspective on Electric Utility Capacity Planning*. Washington, DC: U.S. Government Printing Office.

Conservation Foundation. 1984. *State of the Environment: An Assessment at Mid-Decade*. Washington, DC: The Conservation Foundation.

________________. 1982. *State of the Environment 1982*. Washington, DC: The Conservation Foundation.

EIA (Energy Information Administration). 1984a. U.S. Department of Energy. *Competition and Other Issues in the Natural Gas Market*. DOE/EIA-0489. Washington, DC: U.S. Government Printing Office.

________________. 1984b. U.S. Department of Energy. *Annual Report of Energy Conservation Indicators for 1982*. DOE/EIA-0441. Washington, DC: U.S. Government Printing Office.

________________. 1983. U.S. Department of Energy. *1982 Annual Energy Review*. DOE/EIA-0384. Washington, DC: U.S. Government Printing Office.

Energy Policy Project of the Ford Foundation. 1974. *A Time to Choose: America's Energy Future*. Cambridge, MA: Ballinger Publishing Company.

EPRI (Electric Power Research Institute). 1984. *Effectiveness Review of the Electric Power Research Institute*. Prepared by the

EPRI Board of Directors Effectiveness Review Task Force. Palo Alto, CA: EPRI.

Franssen, Herman T. 1979. *Energy--An Uncertain Future: An Analysis of U.S. and World Energy Projections Through 1990.* A report prepared at the request of the Senate Committee on Energy and Natural Resources, U.S. Congress. Washington, DC: U.S. Government Printing Office.

Glasner, David. 1985. *Politics, Prices and Petroleum: The Political Economy of Energy.* Cambridge, MA: Ballinger Publishing Company.

Gormley, William T., Jr. 1983. *The Politics of Public Utility Regulation.* Pittsburgh, PA: University of Pittsburgh Press.

House Committee on Energy and Commerce. 1982. U.S. Congress. Hearings before the Subcommittee on Fossil and Synthetic Fuels. *Natural Gas Issues,* July 26, 30 and August 6, 9.

Interstate Natural Gas Association of America. 1984. *Natural Gas Carrier Status During the Current Transition: A Critique of Mandatory Contract Carriage.* Policy Analysis 84P-1. Washington, DC: Interstate Natural Gas Association of America.

Landsberg, Hans H. and others. 1979. *Energy: The Next Twenty Years.* Cambridge, MA: Ballinger Publishing Company.

Loftness, Robert L. 1984. *Energy Handbook.* 2nd edition. New York, NY: Van Nostrand Reinhold Company.

Navarro, Peter. 1982. "Our Stake in the Electric Utility's Dilemma." *Harvard Business Review* (May-June): 87-97.

OTA (Office of Technology Assessment). 1983. U.S. Congress. *Industrial Energy Use.* OTA-E-198 (June). Washington, DC: U.S. Government Printing Office.

______________. 1982. U.S. Congress. *Increased Automobile Fuel Efficiency and Synthetic Fuels.* OTA-E-185 (September). Washington, DC: U.S. Government Printing Office.

______________. 1979. U.S. Congress. *Residential Energy Conservation.* OTA-E-92 (July). Washington, DC: U.S. Government Printing Office.

Meier, Alan, Janice Wright and Arthur H. Rosenfeld. 1983. *Supplying Energy Through Greater Efficiency.* Berkeley, CA: University of California Press.

Schelling, Thomas C. 1979. *Thinking Through the Energy Problem.* New York, NY: Committee for Economic Development.

Schurr, Sam H. and others. 1979. *Energy in America's Future: The Choices Before Us.* Baltimore, MD: The Johns Hopkins University Press for Resources for the Future.

Contributors

John Byrne is director of the Center for Energy and Urban Policy Research, and assistant professor of urban affairs and public policy at the University of Delaware. He has conducted research for the U.S. Department of Energy and public and private organizations in the utility sector. He received his doctorate in urban affairs and public policy from the University of Delaware. Dr. Byrne has published extensively in the areas of political economy, energy policy, utility regulation and conservation and renewable energy. He is coeditor of *The Solar Energy Transition: Implementation and Policy Implications, Technology and Energy Choice* and *Energy and Cities.*

Eugene Frankel is a senior analyst and project director at the Office of Technology Assessment (OTA) of the U.S. Congress. Prior to joining OTA, Dr. Frankel served on the professional staff of the Committee on Science and Technology of the U.S. House of Representatives where he was responsible for reviewing and recommending authorization levels for the U.S. Department of Energy's conservation and renewable energy research and development programs. Dr. Frankel holds a doctoral degree in history of science from Princeton University. He is author of a "Multi-year Framework for Solar Energy Research and Development" prepared for the House Science and Technology Committee. He has published numerous articles in the areas of history of science, energy R&D policy, conservation and renewable energy, and technology assessment.

Amory B. Lovins is director of research of the Rocky Mountain Institute, a consultant experimental physicist, and one of the most widely recognized energy analysts of our time. Mr. Lovins has been active in energy policy in more than fifteen countries and has been an advisor to several United Nations agencies and to the U.S. Solar Research Institute and Department of Energy. Educated at Harvard and Oxford Universities, he also has received four honorary doctorates. Among Mr. Lovins' major energy energy publications are: *Soft Energy Paths: Toward a Durable Peace, Non-Nuclear Futures: The Case for an Ethical Energy Strategy, Energy/War: Breaking the Nuclear Link* and *Brittle Power: Energy Strategy for National Security.*

The Honorable Richard L. Ottinger served as U.S. Representative from New York between 1965 and 1970 and again from 1974 through 1984. During his tenure, he chaired one of the most important energy policy-writing bodies in the U.S. Congress, the Energy Conservation and Power Subcommittee of the House Committee on Energy and Commerce. Under his chairmanship the Subcommittee initiated much of the legislation encouraging energy conservation in the U.S. In addition, the Subcommittee enacted the landmark Nuclear Waste Policy Act and developed programs in the areas of utility regulation, regional electric power planning, renewable energy and home energy assistance to low-income families. Mr. Ottinger is currently professor of law at Pace University.

Daniel Rich is senior research associate in the Center for Energy and Urban Policy Research and professor of urban affairs, public policy and political science at the University of Delaware. He received his doctoral degree in political science from the Massachusetts Institute of Technology. Dr. Rich's research has been in the areas of political economy, energy policy, utility regulation, conservation and renewable energy, and urban policy. He is coauthor of *Strategic Alternatives for Energy Conservation* and coeditor of *The Solar Energy Transition: Implementation and Policy Implications, Technology and Energy Choice* and *Energy and Cities.*

J. David Roessner is associate professor in the School of Social Sciences, Georgia Institute of Technology, where he is associated with the Graduate Program in Technology and Science Policy. Previously, he served as principal scientist at the Solar Energy Research Institute, policy analyst with the National Science Foundation's R&D Assessment Program and as an engineer with Hewlett-Packard Co., Inc. Dr. Roessner received his Ph.D. in science, technology and public policy from Case Western Reserve University. In addition to a number of technical reports, he has published in such journals as *Policy Analysis, Policy Sciences,* and *Research Policy.* He is the editor of the forthcoming book, *Government Innovation Policy: Design, Implementation, Evaluation.*

Maxine Savitz is president of the Lighting Research Institute. For sixteen years, Dr. Savitz has been in the forefront of those advocating conservation-based energy options and is an acknowledged leader in energy policy. She has held positions at the National Science Foundation, the Federal Energy Administration, and the Energy Research and Development Administration. From 1979 to January

1983, Dr. Savitz was the Deputy Assistant Secretary for Conservation at the U.S. Department of Energy. She received her doctorate in chemistry from the Massachusetts Institute of Technology. Dr. Savitz is the editor of *Energy & Buildings*.

Richard T. Sylves is associate professor of political science at the University of Delaware. He received his doctorate at the University of Illinois. His publications have appeared in such journals as *The Public Administration Review*, *Publius*, and *The Social Science Quarterly*. Research for his chapter in this volume was conducted at the History Division Archives of the U.S. Department of Energy and is incorporated in his forthcoming book, *America's Nuclear Oracles: A Political History of the AEC General Advisory Committee (1947-1977)*.

Grant P. Thompson, a senior associate at The Conservation Foundation in Washington, D.C., received his legal training at Oxford University and Yale Law School. Active in energy policy research since 1972, he is a past member of the Energy Research Advisory Board of the Department of Energy, and a present member of the Energy Engineering Board of the National Academy of Sciences and the Advisory Council of the Electric Power Research Institute. He has authored or coauthored numerous books and articles, including *Building to Save Energy* and *Energy: The Next Twenty Years*. In addition, he has been responsible for a number of reports on energy research and development for the U.S. Office of Technology Assessment and the U.S. Department of Energy.

For Product Safety Concerns and Information please contact our EU
representative GPSR@taylorandfrancis.com
Taylor & Francis Verlag GmbH, Kaufingerstraße 24, 80331 München, Germany

www.ingramcontent.com/pod-product-compliance
Ingram Content Group UK Ltd.
Pitfield, Milton Keynes, MK11 3LW, UK
UKHW022202190926
475702UK00003B/155

* 9 7 8 1 1 3 8 5 3 7 5 3 8 *